Plasticity of the Central Nervous System

Proceedings of the Second Convention
of the Academia Eurasiana Neurochirurgica
Hakone, October 5–8, 1986

Edited by
Keiji Sano and Shozo Ishii

Acta Neurochirurgica
Supplementum 41

Springer-Verlag Wien New York

Professor Keiji Sano, M.D., D.M.Sc., F.A.C.S. (Hon.)
Department of Neurosurgery, Teikyo University Hospital, Tokyo, Japan

Professor Shozo Ishii, M.D., D.M.Sc.
Department of Neurosurgery, Juntendo University, Tokyo, Japan

With 95 Figures

Library of Congress Cataloging-in-Publication Data. Academia Eurasiana Neurochirurgica. Convention (2nd: 1986: Hakone-machi, Japan) Plasticity of the central nervous system. (Acta neurochirurgica. Supplementum, ISSN 0065-1419: 41). 1. Central nervous system—Regeneration—Congresses. 2. Neuroplasticity—Congresses. 3. Nerves—Growth—Congresses. I. Sano, Keiji. II. Ishii, S. (Shōzō), 1923– . III. Title. IV. Series. [DNLM: 1. Central Nervous System—physiology—congresses. 2. Neuronal Plasticity—congresses. W1 AC8661 no. 41/WL 300 A168 1986p] QP372.6.A255 1986. 612'.82. 87-32222

ISSN 0065-1419
ISBN-13:978-3-7091-8947-4   e-ISBN-13:978-3-7091-8945-0
DOI: 10.1007/978-3-7091-8945-0

Contents

Acta Neurochirurgica, Suppl. 41, 1–2 (1987)

Cerebrum Convalescit

In Memoriam Hans Werner Pia

Keiji Sano

President of the Eurasian Academy, Department of Neurosurgery, Teikyo University Hospital, Itabashi-Ku, Tokyo, Japan

On behalf of the Officers of the Academy, I have the honour and pleasure to give you a cordial welcome to the Second Convention of the Eurasian Academy of Neurological Surgery.

Last year (1985), our first Convention was held in Bonn, Bundesrepublik Deutschland. At that time, Prof. Hans Werner Pia was the first president of this Academy.

It is with great sadness that I must inform you of his decease on July 9th of this year (1986). He was a close personal friend of mine and also a widely respected and renowned colleague. He was one of the founders of this Academy; without his dedicated efforts and inspiration it would not have come into existence. We are greatly indebted to him.

I enjoyed his friendship for more than 20 years. We first met when we both were invited to a professor's house in Copenhagen in 1965 when the 3rd International Congress of Neurological Surgery was being held there. He impressed me as a frank, kind-hearted and amiable man. We had both experienced and survived the War and its misery. We then saw the fall and subsequent rise of our respective fatherlands. We each had our neurosurgical training in a similar difficult postwar situation—he in Germany, I in Tokyo.

We each assumed at about the same time the responsibility of running a neurosurgical department. We also travelled together, with Lore Pia, of course, in various places in Germany. Romantische Strasse, Rheinland, Hessen and many other towns and areas. We shared a deep love of music. I especially cherish the sweet memories of visits to Bayreuth. He kindly invited

me there in two summers. We listened to magnificent Musikspiele of Wagner in an exalted mood, Lohengrin, Tannhäuser, Parsifal, Meistersinger and others.

When we had to create the logo of the Academia, he asked me to coin the motto which should be put into the logo. I replied, "Humanitate et Arte" (with humanity and art). He changed it to "Humanitati et Arti" (to Humanity and to art). He meant by this, this Academia is dedicated to Humanity and to Art, both of which are equally the backbone and the symbol of the Academia. He was really a man who dedicated himself to humanity and art. He, therefore, was himself the symbol of the Academia.

His scientific contributions as well as his superb surgical skill brought him undying international fame. His works on signs and mechanisms of central dysregulation were monumental achievements not only in neurosurgery but in neurosciences in general. He had enormous experience of intracranial aneurysms, cerebral and spinal arteriovenous malformations and intracerebral haematomas. He edited and published monographs about each of these subjects. He had also vast clinical experiences in plasticity of the central nervous system so that he was scheduled to be the first speaker of the scientific sessions of this Convention.

He was an enthusiastic teacher and an ardent promoter of postgraduate training in neurosurgery. The establishment in Europe of postgraduate courses and seminars on various aspects of neurosurgery should be credited mainly to him.

He had many friends and pupils also in this country. All of them respected and loved him. As a consequence, he was unanimously elected Honorary Member of the Japan Neurosurgical Society several years ago.

His untimely death is a great loss not only for European neurosurgery but for Japanese neurosurgeons.

Frau Dr. Lore Pia! Please accept our sincere condolences and heartfelt sympathy. Probably our feeling and sentiment is best expressed by Goethe's Epilog zu Schillers Glocke.

Ach! wie verwirrt solch ein Verlust die Welt!
Ach! was zerstört ein solcher Riß den Seinen!
Nun weint die Welt, und sollten wir nicht weinen?

Denn er war unser! Wie bequem gesellig
Den hohen Mann der gute Tag gezeigt,
Wie bald sein Ernst, anschließend, wohlgefällig,
Zur Wechselrede heiter sich geneigt, ...

His spirit still lives with us. May he rest in peace!

The Leitmotiv of the scientific sessions of the second Conventioin is *Cerebrum convalescit*—literally, "the brain recovers". The focus will be on the *plasticity of the central nervous system*, one of the most favourite subjects of Prof. Pia. We are very glad to have with us many distinguished specialists in this field. It is hoped that ultimately such research can be translated into practical applications in the field of neurological surgery.

Plasticity is related to cellular growth in general or tumour growth in particular. Problems and progress in researches into growth factors and carcinogenesis will be presented and discussed at the last session of the Convention.

Present advances in basic science point the way to future success in neurosurgery. Eventually, research results may, for example, be utilized in curing paralysis and ameliorating other conditions which at present seem irreversible.

This conference symbolizes our commitment not only to preserving and prolonging life but also enhancing the quality of life for our patients.

I would like to thank overseas members who were able to attend—particularly since this year's rapid appreciation of the yen was an added consideration in coming to Japan for this conference. I would also like to thank the session's speakers for their outstanding contributions and also welcome the Japanese professors of neurosurgery who are with us.

The Academy's charter specifies that the purposes of the organization are not only *scientific-academic exchange and mutual understanding*, but it is also to afford *opportunities for social and cultural contact*. To this end, a number of activites have been planned which we hope you will enjoy. We also hope the next few days will enable you to renew old acquaintanceships and forge new friendships.

I sincerely wish you a very pleasant stay in Japan.

Acta Neurochirurgica, Suppl. 41, 3–7 (1987)

Molecular Events Associated with Neural Development

K. Obata, T. Shirao, N. Kojima, and H. Tanaka

Department of Pharmacology, Gunma University School of Medicine, Maebashi, Japan

Summary

Analyses of the chick embryo using two-dimensional gel electrophoresis or immunohistochemistry with monoclonal antibodies (MAbs) revealed several molecules which were highly concentrated in the nervous tissue. The properties and developmental changes of these substances were investigated. Three acidic proteins (molecular weights 95, 100 and 110 kDa, respectively) appeared characteristically during the embryonic development. One of them predominated alternatively at each developmental stage. They showed a high degree of homology in the peptide map and complete cross-reactivity with any specific antibodies. Therefore these proteins were considered to make a protein family and were collectively named drebrins. MAbs were produced against the membrane fraction of the neural tubes and somites of three-day chick embryos. Five MAbs stained the migrating neural crest cells and their progeny in histochemistry. In addition, the otic vesicle, including the differentiated labyrinth, and subtypes of human lymphocytes were recognized by these MAbs.

Keywords: Nervous system-specific protein; development; two-dimensional gel electrophoresis; monoclonal antibody; chick embryo.

Development of the nervous system is brought about by a combination of several fundamental processes such as the proliferation and migration of nerve cells, the directed extension of nerve fibres and the synapse formation on their target cells. In order to disclose the molecular mechanism of the neural development it is necessary to identify any substances which may be involved in such processes and to investigate their molecular nature and physiological roles in the developing nervous system. One approach is to employ such elementary phenomena as cellular adhesion or neurite outgrowth for the *in vitro* assay and to look for active substances in the embryonic material. Adopting this approach Edelman and his collaborators (1984) discovered cell adhesion molecules (CAMs) as a key molecule in the development. Several trophic factors, promoting the neuron survival and the neurite growth, have also been detected in the neural tissue and culture supernatant (Berg 1984, for review).

Another approach is to find out the proteins which are expressed at a limited developmental stage in the restricted region of the developing nervous system and then to explore their physiological function. For detection of such substances, not only biochemical analysis but also monoclonal antibody (MAb) technique is useful. In the present communication we describe two lines of study we have followed adopting the latter approach. One is the identification of a group of developmentally regulated brain proteins, drebrins, with quantitative two-dimensional gel electrophoresis (2 DGE) and the other is the analysis of neural crest-selective antigens with MAbs.

I. Identification and Developmental Changes of Drebrins and Their mRNAs

The avian optic tectum is a uniform and regularly layered structure and develops correctly on timetable, as revealed by Cowan and his colleagues (LaVail and Cowan, 1971 a, b). By 2 DGE of O'Farrell (1975) we systematically surveyed the expression of proteins in the optic tectum at different developmental stages of the chick (Shirao and Obata, 1985). Comparison was made on a pair of electrophoretograms in which the same amounts of tissue were simultaneously electrophoresed and stained with Coomassie brilliant blue. Fifty-four protein spots (S 1–S 54) were numbered (Fig. 1): most of them were found from the beginning (4-day embryo) and remained unchanged until adulthood. There were eight spots that changed their staining intensities remarkably. Four of them (group 1) appeared on the way, then increased and were main-

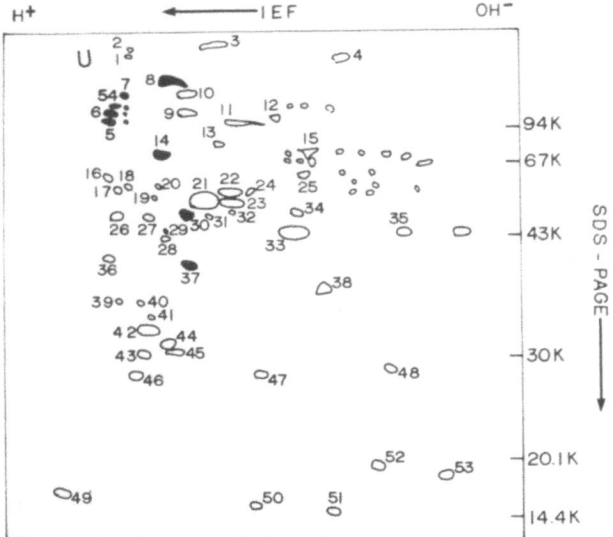

Fig. 1. Composite diagram of major proteins in developing chick optic tectum revealed by two-dimensional gel electrophoresis (2 DGE) which consists of isoelectric focusing (IEF) and sodium dodecyl sulphate-polyacrylamide gel electrophoresis (SDS-PAGE). Out of 54 major protein spots, 8 spots (filled) changed their staining intensities strikingly during development (Shirao and Obata, 1985)

The changes in the relative amount of these three proteins in the optic tectum are shown in Fig. 2. In the dorsal middle portion, E 1 appeared by day 4, reached a maximum level on day 7 and then decreased. E 2 appeared by day 7, reached a maximum level on day 9, retained this level until day 18, and then decreased. Drebrin A appeared by day 15 on Coomassie staining and then increased (with a more sensitive immunoblot, it was detected also in an 11-day embryo, Fig. 4). In this connection, the morphological development of the chick optic tectum (LaVail and Cowan, 1971 a, b) is summarized. Cell proliferation at the ventricular zone reaches the maximum level on embryonic day 6 and then decreases. From day 6 to day 12 the differentiated cells migrate actively to the cortical layers, and the neurons grow their axons extensively from day 6 to day 18.

These proteins were found with similar developmental change also in other parts of the central nervous system. However, the time course varied from area to area: in the cerebellum, which develops late and slowly, drebrin E 1 was still presenmt in a 13-day embryo. At this age E 1 was lost in the optic tectum. Even in the optic tectum the E 1–E 2 conversion and the increase in group 1-proteins occurred earlier in the rostral portion than in the caudal portion. This corresponds to the rostro-caudal gradient of histological development (LaVail and Cowan, 1971 a). In 2 DGE, no form of drebrins was detected in the liver at any developmental stage. Immunoblot analyses of non-neural tissues using the specific antibodies (see below; Shirao and Obata, 1986) confirmed the lack of drebrins in the liver but revealed the presence of E 1 and E 2 drebrins in the skeletal and intestinal muscles of the chick embryos. Thus, drebrin A is specific for the neural tissue but E 1 and E 2 are expressed temporarily in some non-neural cells.

tained in adult chicken. Two proteins (group 2) were already observed in a four-day embryo and then disappeared gradually. The remaining two (S 5 and S 6, group 3) appeared and then disappeared during the course of embryonic development. We concentrated our experiments on these temporarily expressed proteins. As described below, S 5, S 6 and also S 54 in group 1 were intimately related, forming a family. They were collectively named drebrin (developmentally regulated brain protein) and S 5, S 6 and S 54 were called E (embryonic type) 1, E 2 and A (adult type), respectively. Molecular weights of E 1, E 2 and A 2 were 95, 100 and 110 kilodaltons, respectively. They were acidic with isoelectric points around 4.5.

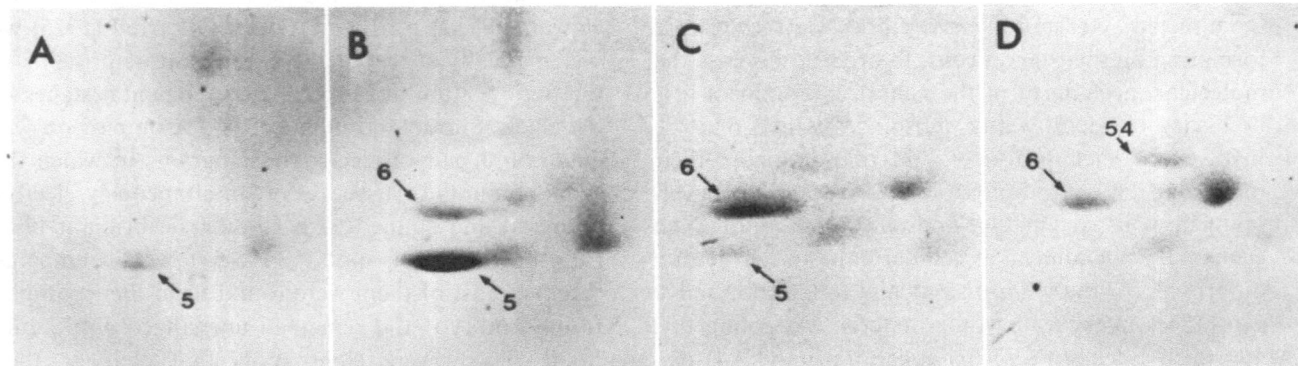

Fig. 2. Developmental changes of drebrins E 1, E 2 and A in the chick optic tectum. Sections of Coomassie brilliant blue-stained electrophoresis gels. A) embryonic day 4; B) day 7; C) day 15; D) newly hatched (Shirao and Obata, 1985)

5

The chemical properties were investigated in the following experiments. E 1 and E 2 were purified from the whole brain of 11-day chick embryos by a procedure consisting of acid (pH 5.5) precipitation, DEAE-Sepharose chromatography, ammonium sulphate (30% saturation) precipitation and SDS-polyacrylamide gel electrophoresis (SDS-PAGE). Fig. 3 A shows homogeneity of E 1 and E 2 thus obtained. Purified E 1 and E 2 drebrins were compared by gel electrophoresis peptide mapping using *S. aureus* V 8 protease. They had at least 11 common peptides and six distinct ones (Fig. 3 B). This indicates a strong degree of structural homology between these two forms. Using purified E 1 and E 2 as immunogen, we tried to produce antibodies specific for either form. Anti-E 1 antiserum, anti-E 2 antiserum and 5 independent MAbs were obtained. All antibodies specifically recognized three forms of drebrins (Fig. 4): none of these antibodies were specific for one of three drebrins. Immunological analyses indicated that four MAbs recognized different epitopes of the same molecules. Cross-reactivity of all antibodies obtained indicates that three forms of drebrins are closely related proteins. In order to

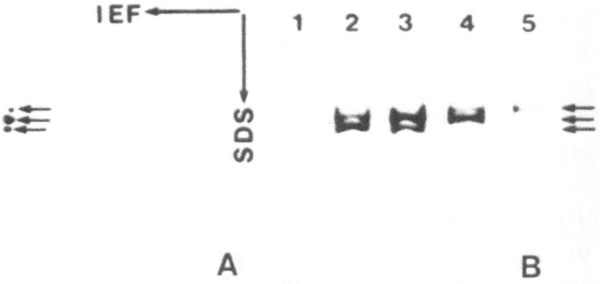

Fig. 4. Immunoblot analysis of chick optic tectum with an anti-drebrin monoclonal antibody (MAb M2F6). A) 2 DGE of 11-day chick embryo proteins. B) SDS-PAGE. Lane *1* embryonic day 4; *2* day 7; *3* day 9; *4* day 16; *5* 10 days postpatching. The arrows indicate E 1, E 2 and A from the bottom (Shirao and Obata, 1986)

elucidate further the relationship between these molecules we have attempted molecular cloning (Kojima et al., in preparation). Poly(A)$^+$ RNAs were obtained from the chick embryo and the corresponding cDNAs were prepared. They were introduced into an expression vector, λ gt 11 and amplified to obtain a cDNA library. Fusion proteins encoded by the recombinant DNA were screened for specific binding

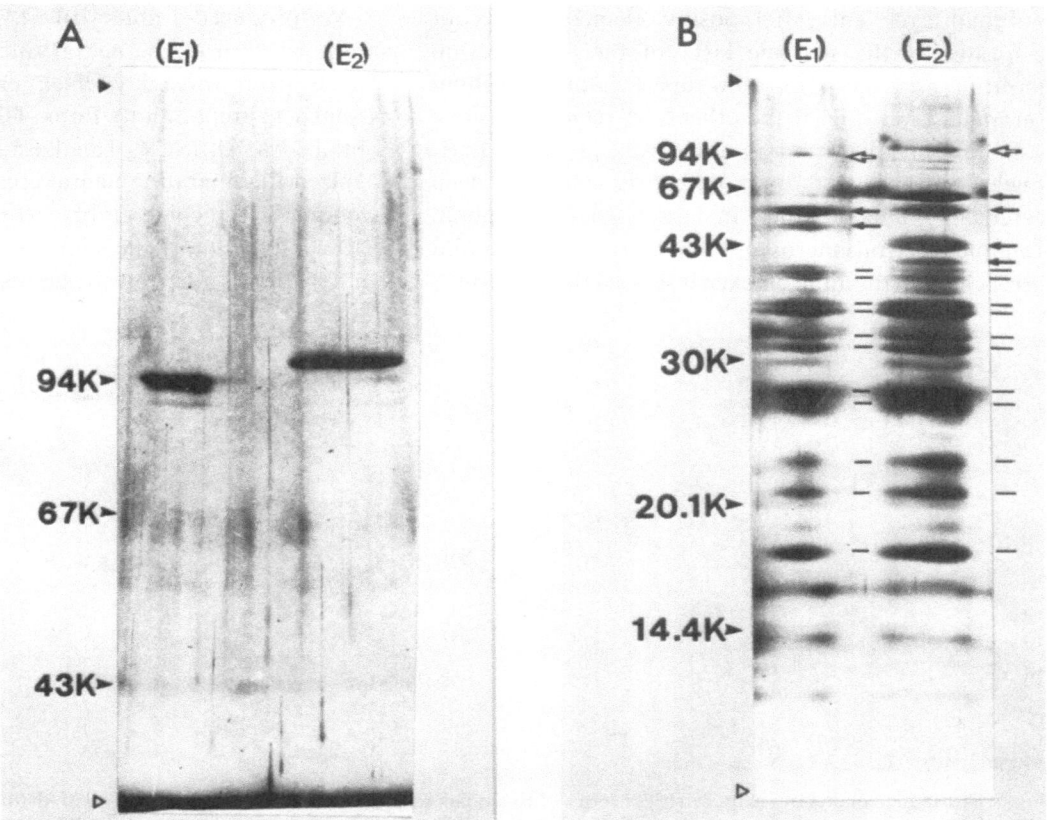

Fig. 3. Purified E 1 and E 2 drebrins (A) and their peptide mapping (B). SDS-PAGE. Silver stain. In B) each protein was treated with *S. aureus* V 8 protease. Open arrows, intact proteins; closed arrows, distinct peptides; lines, common peptides (Shirao and Obata, 1985)

with anti-drebrin antibodies. Thus, a cDNA clone, named gDcw 1 was obtained which was specific for drebrin with an insert of 1.4 kb. Messenger RNA was obtained from 7-, 11- and 17-day chick embryos and electrophoresed for Northern blotting. gDcw 1 hybridized with two or three mRNA bands in each preparation. This indicates that three drebrin forms are derived from the corresponding mRNA species but not produced by post-translational modification of a single mRNA product.

Cellular and subcellular localization of drebrins was investigated in the embryonic and adult chick by immunohistochemistry (Shirao and Obata, 1986) and immunoelectron microscopy (Shirao et al., 1987). First immunoreactivity appeared in the cells of the outermost layer in the optic tectum around the fourth day of incubation and in the cells of the premigratory zone of the outer granular layer (Altman, 1972) in the cerebellum (Fig. 5). Neurons in the surface of the optic tectum at this stage have just finished migration from the opposite ventricular zone and are extending axons of the tectobulbar tract. Neurons in the cerebellar premigratory zone have finished proliferation at the proliferative zone and begin to migrate into the inner granular layer. Thereafter, positive elements were concentrated in the synaptic layer of the optic tectum (stratum griseum et fibrosum superficiale), cerebellum (granular layer, Fig. 5) and other CNS regions. Positive somata were dispersed in some central nuclei (*e.g.*, nucleus isthmi pars parvocellularis and the deep cerebellar nuclei) and in the spinal ganglion. Immunoelectron microscopy of the optic tectum and cerebellum in the adult chicken disclosed that drebrin A

was confined in postsynaptic region of the dendrites but not found in presynaptic axons, neuronal somata or glial elements. Immunoreactivity was distributed in the cytoplasm and not associated with membrane or organellae.

The time course of expression and the distribution in just differentiated neurons and postsynaptic dendrites suggest that drebrins play a role in the extension of neurites and the formation, maintenance and remodelling of synapses.

II. Common Antigenicity in the Neural Crest and Otic Vesicle

The peripheral nervous system arises from the neural crest and the ectodermal placodes. Migration of the neural crest cells and their differentiation into various types of cells is one of the major subjects in developmental neurobiology (Le Douarin, 1986). Recently two MAbs, HNK-1 (Tucker et al., 1984) and NC-1 (Vincent and Thiery, 1984), were demonstrated to label the neural crest cells specifically. Specific cell markers are very useful in studying such dynamically changing tissues as the neural crest.

We produced a library of MAbs against the membrane fraction of the neural tube-somite regions obtained from three-day chick embryos (Obata and Tanaka, in preparation). Four of them and a previously obtained MAb (SC 4, Tanaka and Obata, 1984) recognized the migrating neural crest cells on the frozen sections of the chick embryo (Fig. 6 A). The staining pattern was almost the same as reported for HNK-1 and NC-1. In addition to the neural crest cells, these

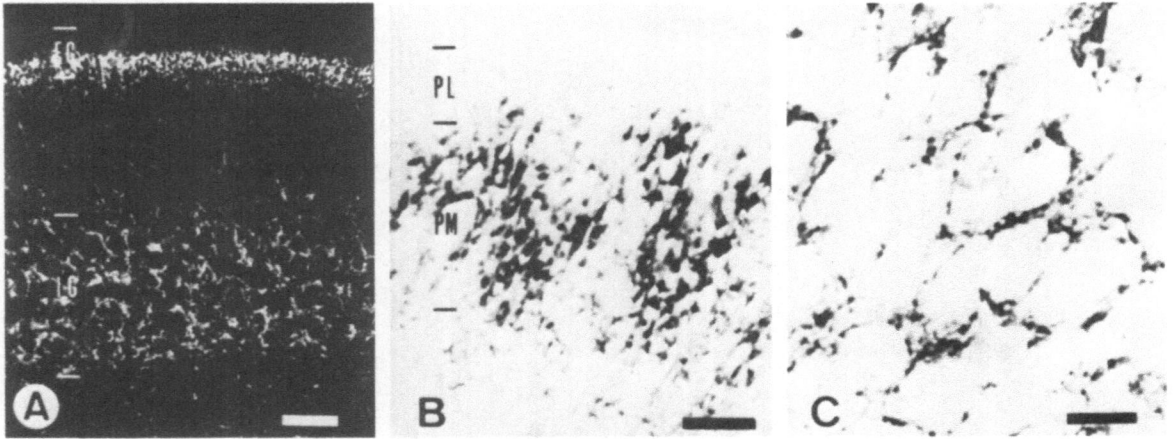

Fig. 5. Distribution of drebrins in the cerebellar cortex of 16-day chick embryo. Immunohistochemistry with MAb M2F6 followed by treatment with the FITC-conjugated (A) and HRP-conjugated second antibody (B, C). In the external granular layer (*EG*), premigratory zone (*PM*) was labelled but proliferative zone (*PL*) was unstained. Internal granular layer (*IG*) was also labelled. Higher magnification of IG in C shows that cell processes but no cell somata were stained. Bars represent 40 µm in A) and 10 µm in B) and C) (Shirao and Obata, 1986)

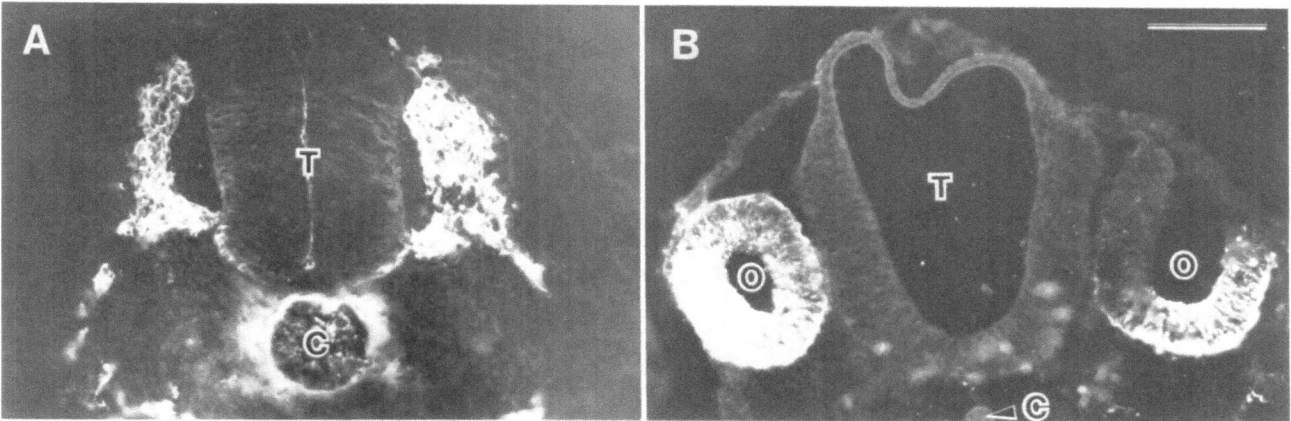

Fig. 6. Immunostaining with an anti-neural crest monoclonal antibody. A) trunk level of 3-day chick embryo. B) hind-brain level of 2-day chick embryo. *T* neural tube; *C* notochord; *O* otic vesicle. In A) the diffuse reactivity was observed around the notochord too. Bar, 100 μm

MAbs stained the primordium (otic placode and otic vesicle) and membrane labyrinth of the inner ear (Fig. 6 B). Sensory neurons in the acoustic ganglion derive from the otic vesicle. On the other hand, the satellite cells in the same ganglion derive from the neural crest. It is not known why the otic vesicle and the neural crest have the common antigen but it might be possible that the common substance on the surface of the cells of different origin will serve for mutual recognition to form a single tissue.

Six MAbs appeared to bring high molecular-weight glycoprotein but did not recognize the same epitope. Periodate treatment of the embryonic tissue ablished the antigenicity of only three MAbs. SC 4 stained human B lymphocyte lines (Raji and U 937) but not a T-cell line, Molt 4. On the other hand, the other five MAbs stained about 25% of Molt 4 cells as in the case of HNK-1 (Abo and Balch, 1981). It is to be determined what is the role of these antigens common to the neural tissue and lymphocytes.

References

1. Abo T, Balch CM (1981) A differentiation antigen of human NK and K cells identified by a monoclonal antibody (HNK-1). J Immunol 127: 1024–1029

2. Altman J (1972) Postnatal development of the cerebellar cortex in the rat. I. The external germinal layer and the transitional molecular layer. J Comp Neurol 145: 353–398

3. Berg DK (1984) New neuronal growth factors. Ann Rev Neurosci 7: 149–170

4. Edelman GM (1984) Modulation of cell adhesion during induction, histogenesis, and perinatal development of the nervous system. Ann Rev Neurosci 7: 339–377

5. LaVail JH, Cowan WM (1971 a) The development of the chick optic tectum. I. Normal morphology and cytoarchitectonic development. Brain Res 28: 391–419

6. LaVail JH, Cowan WM (1971 b) The development of the chick optic tectum. II. Autoradiographic studies. Brain Res 28: 421–441

7. LeDouarin NM (1986) Cell line segregation during peripheral nervous system ontogeny. Science 231: 1515–1522

8. O'Farrell PH (1975) High resolution two-dimensional electrophoresis of proteins. J Biol Chem 250: 4007–4021

9. Shirao T, Obata K (1985) Two acidic proteins associated with brain development in chick embryo. J Neurochem 44: 1210–1216

10. Shirao T, Obata K (1986) Immunochemical homology of three developmentally regulated brain proteins and their developmental change in neuronal distribution. Dev Brain Res 29: 233–244

11. Shirao T, Inoue HK, Kano Y et al (1987) Localization of a developmentally regulated neuron-specific protein S 54 in dendrites as revealed by immunoelectron microscopy. Brain Res 413: 374–378

12. Tanaka H, Obata K (1984) Developmental changes in unique cell surface antigens of chick embryo spinal motoneurons and ganglion cells. Dev Biol 106: 26–37

13. Tucker GC, Aoyama H, Lipinski M et al (1984) Identical reactivity of monoclonal antibodies HNK-1 and NC-1: Conservation in vertebrates on cells derived from neural primordium and on some leucocytes. Cell Differ 14: 223–230

14. Vincent M, Thiery J-P (1984) A cell surface marker for neural crest and placodal cells: Further evolution in peripheral and central nervous system. Dev Biol 103: 468–481

15. Young RA, Davis RW (1983) Efficient isolation of genes by using antibody probes. Proc Natl Acad Sci USA 80: 1194–1198

Correspondence: Dr. Kunihiko Obata, Department of Pharmacology, Gunma University School of Medicine, 3-39-22 Showamachi, Maebashi 371, Japan.

Acta Neurochirurgica, Suppl. 41, 8–17 (1987)

Regeneration of the Cerebellofugal Projection After Transection of the Superior Cerebellar Peduncle in the Cat

S. Kawaguchi

Department of Physiology, Institute for Brain Research, Faculty of Medicine, Kyoto University, Kyoto, Japan

Summary

In contrast to the current concept of abortive regeneration of mammalian central axons, the occurrence of marked, functionally active, regeneration of the cerebellofugal projection was proved in the cat after complete transection of the decussation of the brachium conjunctivum (BCX). Because the BCX is a complete crossing it was transected completely by pushing down an edged U-shaped wire to the base of the brain stem in the midline, and the wire was left *in situ* to mark the lesion. Later, horseradish peroxidase was injected into the cerebellar lateral and interpositus nuclei to label the cerebellofugal projection arising from these nuclei; axonal regeneration was proved by demonstration of labelled fibres passing through the area enclosed by the U-shaped wire. By this procedure the origin, course, and destination of the regenerated fibres were identified unambiguously. Most of the regenerated axons took a course similar to that of the normal projection and terminated in the normal projection areas, whereas a small proportion of fibres showed an aberrant course and termination. Functional connectivity of the regenerated cerebello-thalamic projection was tested electrophysiologically in the same animals examined morphologically. In all animals in which marked axonal regeneration occurred, cerebellocerebral responses, as in intact animals, were evoked in the frontal motor and parietal associated cortices. Study of the time-course of regeneration revealed that the cut ends of axons began to swell as early as 15 min after transection, produced terminals tipped by growth cones in 14–24 hours, grew to cross the lesion in 3 days, and distributed dense terminals in the thalamus by 19 days.

Keywords: Axonal regeneration; cerebellofugal projection; functional recovery, cat.

Ramón y Cajal who achieved a monumental work on degeneration and regeneration of the nervous system concluded in his book[4] as "Once the development was ended, the founts of growth and regeneration of the axons and dendrites dried up irrevocably. In adult centres the nerve paths are something fixed, ended, immutable." Since then, confirmed by many other investigators, it has been widely accepted that axonal regneration in the mammalian central nervous system is poor, if any, despite the presence of initial signs of regrowth, *i.e.*, growth cones near lesions[5, 8, 20, 24]. Growth cones sprouting from cut axons generally retract without reaching the proper targets and eventually degenerate. This phenomenon is known as abortive regeneration. Studies of recent years, however, have provided evidence that mammalian central axons do not necessarily result in abortive regeneration but can regenerate successfully under certain circumstances[1, 3, 6, 7, 9–11, 21, 22].

It is self-evident that axonal regeneration can be accepted only when the axons under consideration are proved to originate from cut axons but not from neighbouring spared axons. It is not so easy however to prove it. In fact, many papers describing axonal regeneration lack this proof in a strict sense. Complete transection of a central fibre bundle is rather difficult even when a razor-sharp knife is used, and spared fibres tend to be dislocated remarkably[15]. Thus, the presence of fibres coursing in aberrant pathways is not evidence for axonal regeneration. The presence of newly formed fibres or growth cones itself, again, is not evidence for axonal regeneration because collateral sprouting from intact fibres has been well-known to occur after lesions[23]. Furthermore, axonal regeneration must be distinguished from axonal generation, *i.e.*, developmental growth of immature axons which arrived late after the lesion, when the brain of developing animals, particularly of embryos, is used.

The cerebellofugal projection in kittens[12, 15] as well as in adult cats[13] can regenerate markedly, though infrequently, after transection of the decussation of the brachium conjunctivum. For this, conclusive evidence was provided by unambiguous identification of the origin, course, and destination of the axons which were examined by anterograde axonal transport of horseradish peroxidase. Functional connectivity of the regenerated projection was tested and time-course of axonal regeneration was also pursued.

Course and Destination of the Cerebellofugal Projection Arising from the Interpositus and Lateral Nuclei in Intact Animals

Fibres arising from the lateral and interpositus nuclei of the cerebellum form the superior cerebellar peduncle, which is a compact fibre bundle. The entire course of this projection can be visualized up to the terminals by labelling of horseradish peroxidase (HRP) that was injected into these nuclei. As shown in Fig. 1 in horizontal (A) and frontal sections (B) of the brain stem, the superior cerebellar peduncle in intact animals decussates completely at the decussation of the brachium conjunctivum (BCX). The fibres cross the midline mostly at the level of the interpeduncular nucleus and partly at a level rostral and caudal to that nucleus. It is worth noting that uncrossed fibres can never be seen in any sections cut rostral, caudal, dorsal or ventral to the major crossing. While fibres course closely around the interpeduncular nucleus, none of them enters the nucleus. A large part of the crossed fibres turns rostrally almost rectangularly to form the ascending limb and a relatively small number of the crossed fibres turn caudally to form the descending limb of the superior cerebellar peduncle. The ascending limb traverses the brain stem to enter and surround the contralateral red nucleus (Fig. 1 A). These fibres give dense terminals to the red nucleus (RN), and then, on the way projecting onto the thalamus, sparse terminals to the periaqueductal grey (PAG), interstitial nucleus of Cajal (Caj), nucleus of Darkschewitsch (Da), nucleus of posterior commissure (NPC), superior colliculus (SC), pretectum (PRT), Forel's field (FF), and zona incerta (ZI). The bulk of the ascending limb terminates in the thalamus, most densely in the ventral anterior (VA) and ventral lateral (VL) nuclear complex, moderately in the central lateral (CL) nucleus, and sparsely in the lateral posterior (LP) and pulvinar (Pul) complex, centromedian (CM), paracentral (Pc), and central medial (NCM) nuclei. A small number of fibres cross the midline within the thalamus forming the recrossed ispilateral cerebellothalamic projection and terminate sparsely in the ipsilateral VA-VL nuclear complex, CL, CM, Pc, and NCM nuclei. The descending limb of the superior cerebellar peduncle terminates in the contralateral pontine tegmental nucleus (PTN) and the inferior olive (IO). The course of the labelled fibres and distribution of their terminals remain unchanged from birth to adulthood, although the fibres are myelinated in adults but unmyelinated in neonates. Swollen endings described as growth cones can never be observed along the entire course of the superior cerebellar peduncle in kittens from birth. This indicates a prenatal occurrence of growth and termination of the cerebellofugal projection in the cat, which is evidenced also by evoked response[16] and retrograde labelling of cerebellothalamic neurons[18].

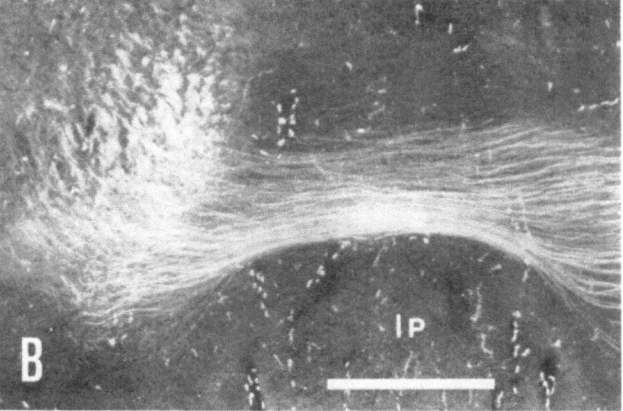

Fig. 1. Dark-field photomicrographs of the superior cerebellar peduncle labelled with HRP in intact animals. A horizontal section (A) and a frontal section (B) of the brain stem; the former section was taken from a six-day-old kitten and the latter section from a one-day-old kitten in which HRP was injected into the cerebellar nuclei immediately after birth. Note that there are no uncrossed fibres, no axons tipped by swollen endings suspectible of growth cones, and no fibres entering the interpeduncular nucleus or the red nucleus ipsilateral to the site of HRP injection. Bar = 500 μm, applicable to A) and B)

Morphological Evidence for Regeneration of the Cerebellofugal Projection

A complete transection of the BCX was accomplished in kittens of various ages and adult cats by implantation of an edged U-shaped trilateral tungsten wire which was pushed down to the base of the brain stem in the midline and left *in situ* (Fig. 2). In these animals HRP was injected into the interpositus and lateral nuclei of the cerebellum at various post-lesion days to label anterogradely the cerebellofugal projection arising from these nuclei. Axonal regeneration can be proved provided fibres labelled with HRP are demonstrated to cross the area enclosed by the trilateral tungsten wire because such fibres cannot be ascribed to collateral sprouting from spared axons or axonal generation in ontogenesis on the basis of findings in the control animals. This is in fact what was demonstrated in eight out of 82 animals examined at 3 to 126 days after the lesion[15] and in three out of 19 adult cats at 25 to 66 days after the lesion[13]. In the following decriptions, the animal age is given as A (days), the post-lesion time is given as B (days), and the animal is identified as A PO B.

An example of the marked axonal regeneration is shown in Fig. 3 which is a horizontal section of the brain stem taken from a kitten (6 PO 19). Many fibres labelled with HRP crossed the previously transected sagittal plane and turned rostralward to ascend the brain stem to enter the red nucleus, just like normal projection fibres. However, there were some aberrant projections in the tegmental field rostral to the BCX

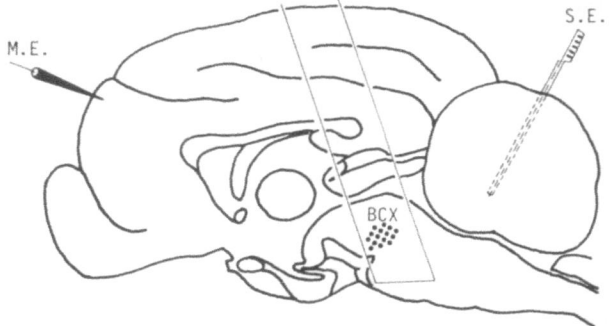

Fig. 2. Experimental design to prove axonal regeneration. A U-shaped trilateral tungsten wire of 200 μm in diameter, a razor-sharp edged bottom side connected with arms of two parallel lateral sides, was used for BCX transection. Complete transection of the BCX was accomplished by implantation of the cutting device which was pushed down in the midline to reach the base of the brain stem and left *in situ* to mark the lesion. Later, HRP was injected into the cerebellar nuclei to demonstrate regenerated fibres passing through the area enclosed by the implanted wire. Cerebellocerebral responses were recorded to test the functional connectivity of regenerated fibres in the same animal for the histological examination

(thin arrow). The population of the regenerated fibres appeared at least half of that of the normal projection fibres although it was not measured quantitatively. The extent of the tissue initially divided by an edged wire was identified by two holes in the horizontal plane which had been produced by the arms of the cutting device. There were dense glial scars in the rostral and caudal portions of the lesion, whereas such scars were absent in and around the area crossed by the regenerated fibres. Because of the absence of demarcated

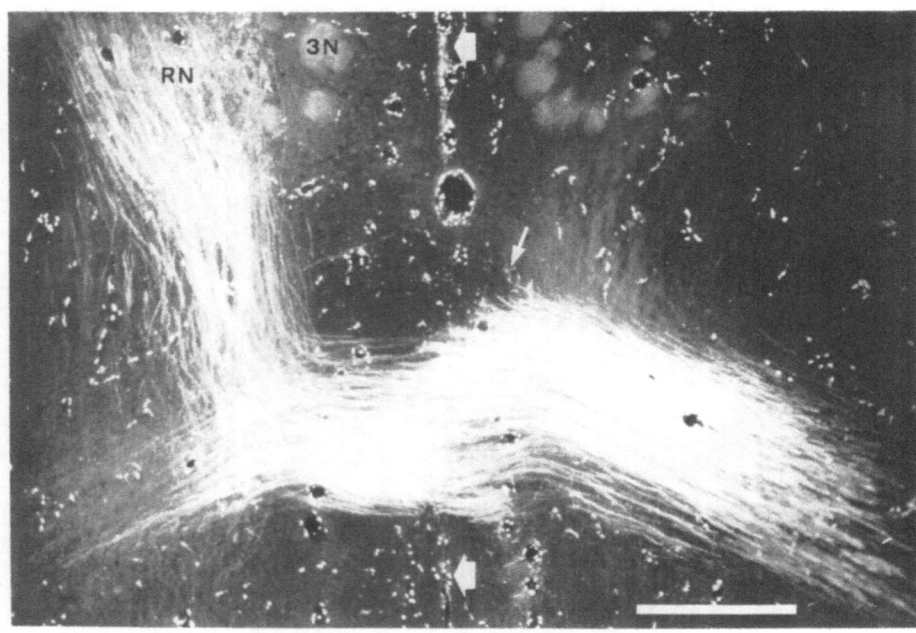

Fig. 3. Dark-field photomicrograph of regenerated superior cerebellar peduncle, taken from a horizontal section of the brain stem (6 PO 19). Aberrant terminals are seen in the tegmental field rostral to the BCX (thin arrow). There were dense glial scars in the rostral and caudal portions (thick arrows) but not in the area of fibre crossing. Bar = 500 μm

glial scars it would be difficult to believe that the area of fibre crossing was initially divided unless the cutting device had been implanted.

Another example of the marked regeneration is shown in Fig. 4 which is a frontal section of the brain stem taken from a kitten (14 PO 33). In this case, a portion of the labelled fibres crossed the midline to enter the contralateral brain stem but some did not cross and concentrated at the midline where the tissue was initially divided. There were dense glial scars above and below the area crossed by labelled fibres (thick arrows), whereas gliosis in the area of crossing was again diffuse and not dense. The crossing fibres were much more dense and twisted as compared with those in control animals. Aberrant projections were seen in the interpeduncular nucleus (thin arrow).

In two of the eight kittens which showed marked regeneration of the crossed contralateral cerebellofugal projection and seven other kittens which did not show such regeneration, a considerable amount of uncrossed ipsilateral cerebellofugal projection was observed, as shown in Fig. 5. This projection might have arisen from the fastigial nucleus which is well-known as sending the uncrossed ipsilateral cerebellofugal projection. This

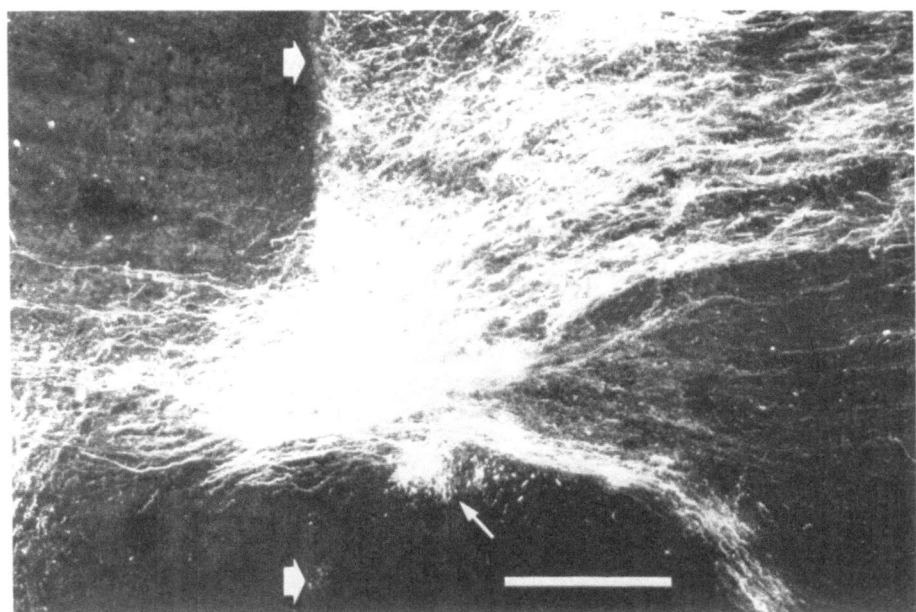

Fig. 4. Dark-field photomicrograph of regenerated superior cerebellar peduncle, taken from a frontal section of the brain stem (14 PO 33). Aberrant terminals were distributed in the interpeduncular nucleus (thin arrow). There were glial scars in the midline dorsal and ventral to the area of fibre crossing (thick arrows). Bar = 500 μm

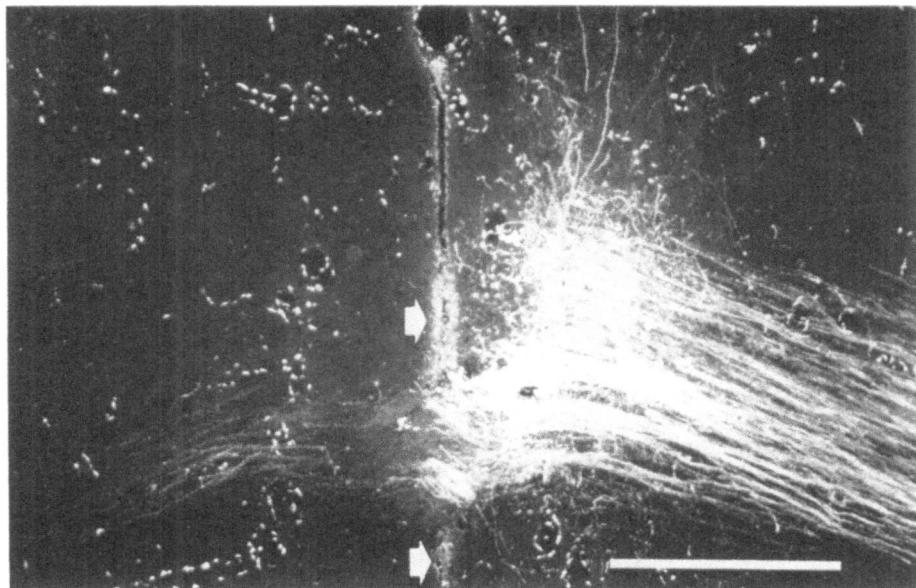

Fig. 5. Dark-field photomicrograph showing aberrant cerebellofugal projection in a horizontal section of the brain stem. A considerable number of fibres remained uncrossed to ascend the ipsilateral brain stem up to the thalamus. Aberrant terminals are seen in the tegmental field surrounding the superior cerebellar peduncle. The section is more dorsal than that in Fig. 3, taken from the same kitten (6 PO 19). Bar = 500μm

possibility, however, appears unlikely for the following reasons. (1) The fastigial nucleus was virtually free from spread of the injected HRP. (2) A similar injection of the same amount of the enzyme in control animals labelled no uncrossed ipsilateral projection. (3) The terminal distribution of this projection differed from that of the fastigiofugal projection. The uncrossed ipsilateral cerebellofugal projection in the operated animals is, therefore, aberrant and considered to have been regenerated because there were no spared axons to sprout collaterals.

In two of the remaining 67 kittens, labelled fibres were observed to make "detours", *i.e.*, the fibres bent ventrally and coursed around the bottom side of the cutting device, and then, bent dorsally. Fig. 6 illustrates an example of the "detours" which were disrupted by removal of the implanted wire from the ventral side. These fibres appeared most likely to be due to axonal regeneration and very unlikely to be due to persistence of initially deflected fibres because a razor-sharp wire had been pushed down to reach the base of the brain stem and a similar procedure in acute experiments transected the BCX completely.

The consequence of BCX transection in the vast majority of animals was failure of marked axonal regeneration. In a large proportion of such animals, however, there were trace amounts of aberrant projection, *i.e.*, the uncrossed ipsilateral cerebellofugal projection which was considered to consist of the regenerated fibres.

In all the cases in which marked axonal regeneration occurred, the initially divided tissue, at least partly, became continuous without producing dense glial scars; and regenerated fibres crossed the midline always through such an area. By contrast, in all the animals in which axonal regeneration was not remarkable or virtually absent, dense glial scars were formed over the entire extent of the lesion. In some of them cavities walled by dense glial and collagenous scars were formed in the lesion. These cavities were presumably due to bleeding at the time of operation although this could not be verified retrospectively.

Course and Destination of the Regenerated Cerebellofugal Projection

Fig. 7 illustrates schematically the course and destination of the regenerated fibres which were largely similar to those of normal projection fibres (filled lines and arrows) but partly aberrant (open lines and

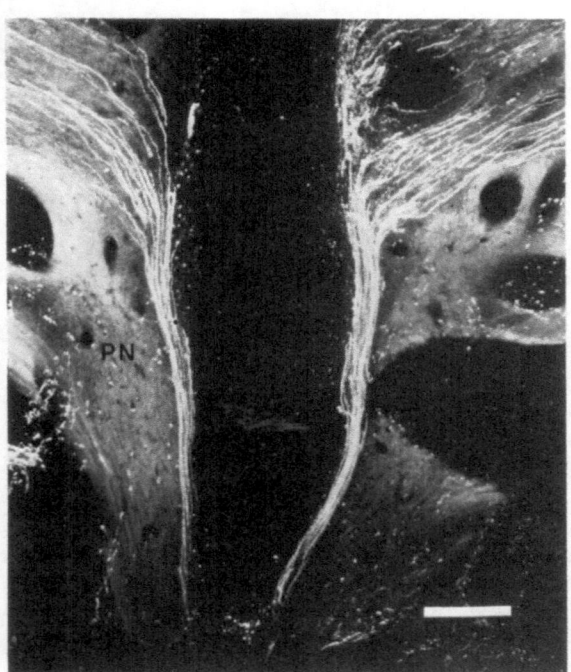

Fig. 6. Dark-field photomicrograph showing axons making detours, taken from a frontal section of the brain stem (32 PO 26). Fibres making detours to bypass the bottom side of a cutting device were disrupted when the device was removed from the ventral side of the pons. The bottom side of the device had reached the base of the brain stem at the time of transection, which was translucently visible underneath the dura mater covering the pons at the time of brain cutting. Bar = 200 µm

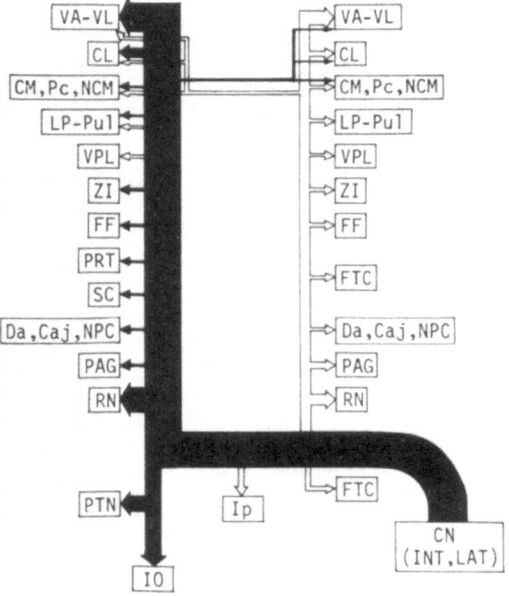

Fig. 7. Schematic illustration of the course and destination of normal and aberrant projections arising from the interpositus and lateral nuclei of the cerebellum. Filled lines and arrows indicate normal projections in the intact as well as in the operated animals. Open lines and arrows indicate aberrant projections in the operated animals

arrows). Fibres either crossing the lesion or making "detours" separated into two bundles in a manner similar to the normal projection fibres: one consisted of a large proportion of fibres which turned rostralward to form the ascending limb and the other consisted of a small proportion of fibres which turned caudalward to form the descending limb. The ascending limb, as in control animals, gave the bulk of its terminals to the contralateral red nucleus and thalamic VA-VL nuclear complex, a relatively small number of terminals to the thalamic CL nuclei, and much smaller number of terminals to the periaqueductal grey, interstitial nucleus of Cajal, nucleus of Darkschewitsch, nucleus of posterior commissure, superior colliculus, pretectum, Forel's field, zona incerta, and several thalamic nuclei (LP-Pul, CM, Pc, and NCM). There were a small number of fibres crossing within the thalamus to form the recrossed ipsilateral cerebellothalamic projection terminating in the ipsilateral thalamic VA-VL nuclear complex, CL, CM, and Pc nuclei.

In nine out of the 82 animals, as shown in Fig. 5, there was a considerable amount of uncrossed ipsilateral cerebellofugal projection arising from the lateral and interpositus nuclei, which were never observed in control animals. The destinations of this projection illustrated in Fig. 7 as open arrows were mostly homologous structures of the destinations of the crossed contralateral cerebellofugal projection in intact as well as in the operated animals. Besides these fibres projecting onto homologous structures of the normal projection, there were another fibres terminating in entirely anomalous destinations such as the VPL, caudal part of the LP-Pul, central tegmental field surrounding the BCX, and interpeduncular nucleus (Fig. 4). The proportion of fibres terminating in the normal destinations was much larger than that of fibres terminating in the aberrant destinations; and for the latter, the proportion of fibres terminating in the homologous structures of the normal projection was much larger than that of fibres terminating in the entirely anomalous destinations.

Time-course of Axonal Regeneration

Morphological changes in axons were examined at various times after BCX transection. For examination of changes within 24 hours of operation, animals were injected with HRP prior to the transection. The order of the procedures was reversed for examination of changes of more than 24 hours after the transection. As early as 15 minutes after transection, some proximal and distal stumps of the cut axons began to swell, and the swelling became increasingly remarkable over a distance of $200 \mu m$ by 30 minutes. In 14 hours all fibre endings showed polymorphic swelling near the lesion: some were presumably growth cones and some were presumably retraction bulbs. The typical endings which were presumed growth cones were large, ramified, and very densely labelled, whereas the typical endings which were presumed retraction bulbs were small, round, and faintly labelled. Besides these typical ones, there were numerous endings which could not be placed under the two categories; they were smooth or spinous club-shaped endings with or without dense labelling.

At 24 hours after transection the presumed growth cones and club-shaped endings were scattered not only near the lesion but also far behind the lesion, and the presumed retraction bulbs faintly labelled with HRP were found 500 to $1,000 \mu m$ behind the lesion. Swelling of distal stumps changed into retraction bulbs which were labelled faintly. In three days a small number of axons tipped by a swollen ending crossed the midline where the BCX had been transected. After seven days, such axons increased in number and ascended through the contralateral brain stem to reach the red nucleus. Swollen endings, presumably growth cones, were observed sporadically not only along the way to the red nucleus but also near the lesion and behind the lesion at more than a millimetre. In the kitten (6 PO 19) shown in Fig. 3, dense terminals were distributed in the contralateral red nucleus and thalamic VA-VL nuclear complex, which indicates that a large number of regenerated axons can re-establish their terminal distribution within three weeks. This does not necessarily mean that the regenerative process ceases. Actually, it continues for a much longer period of time since swollen endings, presumably growth cones, were observed in various portions of the brain stem and in the thalamic VA-VL nuclear complex more than four weeks after the operation.

For the majority of animals in which successful regeneration of axons did not occur, labelled fibres were virtually absent two to three weeks after the operation in agreement with the well-documented abortive regeneration[4]. However, this was not always the case. In the minority of animals a considerable number of fibres were persistently labelled four weeks or more after the operation even though marked axonal regeneration did not occur. In one animal examined at 125 days after being operated on at 31 days old, a significant number of swollen endings resembling growth cones were labelled near the lesion in addition to numerous fibres facing glial scars and a small number of fibres ascending the brain stem ipsilaterally.

Regarding the neoformation of axon terminal, Ramón y Cajal proposed the concept of "direct or distal sprouting" and "indirect or proximal sprouting": the former occurs at the end of axons whereas the latter occurs at a certain proximal segment of the axon when the distal portion is necrotic[4]. This distinction is not absolute because all axons lose some protoplasm as a result of direct injury. The formation of uncrossed ipsilateral cerebellofugal projections shown in Fig. 5 are accounted for by axonal neoformation due to "indirect or proximal sprouting". Axons tipped by retraction bulbs, therefore, do not necessarily mean that they are under the down-hill degenerative process; but rather, they may be turned into regenerating axons. On the other hand, axons tipped by regenerating endings will finally be turned into degenerating ones unless they reach target neurons. Thus, it appears very likely that the regenerative and degenerative processes are concomitant in each injured axon and that whether the axon regenerates or degenerates depends on which process overwhelmed the other, as in a tug-of-war.

Exactly similar to the terminals of normal projection fibres, most of the terminals of regenerated axons were very fine. However, in the operated animals but not in the control animals, various forms of much coarser terminals, very likely the developing terminals, were observed in the thalamus. Polymorphism of these terminals appears to correspond to the different stages of development of growing tips of regenerated axons. Because of their gradual changes in the grade of metamorphosis, it is easy to arrange them in the presumptive order of development as depicted in Fig. 8 from A to E. All of these illustrations were taken from thalamic sections of one and the same kitten (6 PO 19). The vast majority of terminals were fine; and coarser terminals, particularly those depicted in A and B, were small in number, sporadically seen around the fine terminals as shown in F.

Functional Connectivity of Regenerated Cerebellofugal Projection

Stimulation of the interpositus or the lateral nucleus of the cerebellum in intact animals induces contralaterally

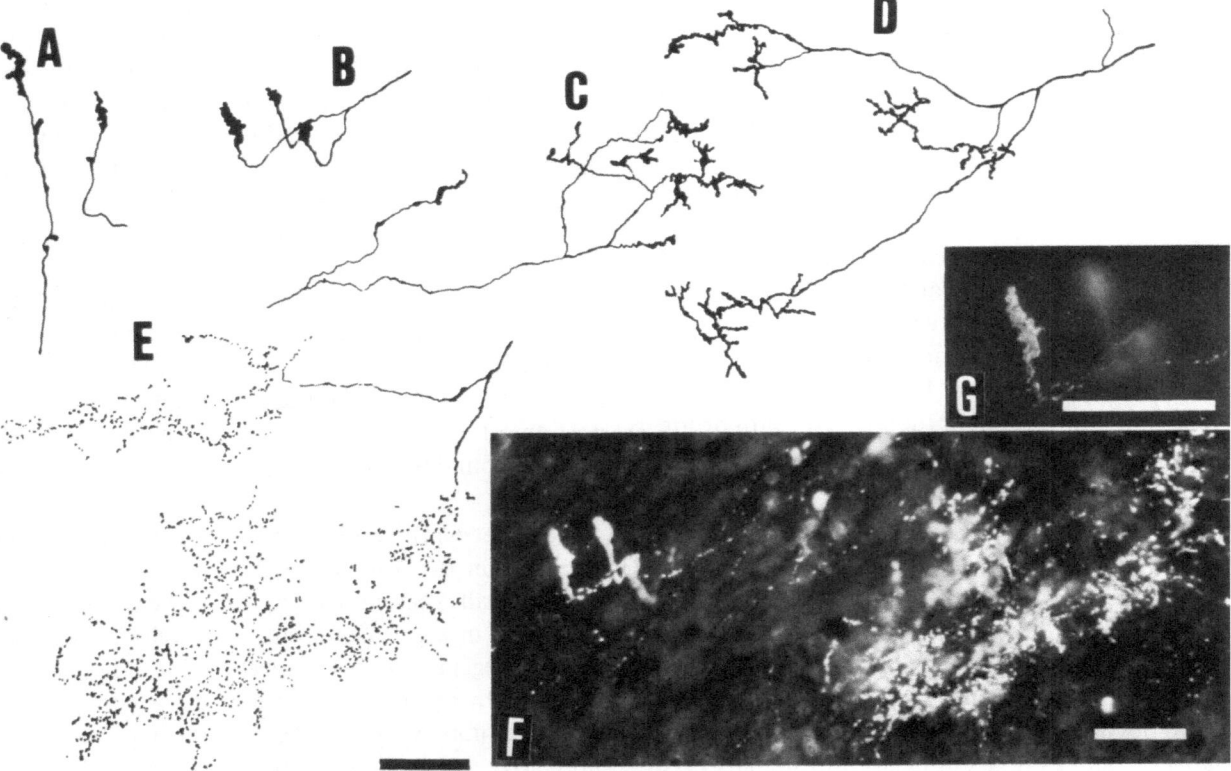

Fig. 8. Line drawings and dark-field photomicrographs of polymorphic terminals of regenerated cerebellothalamic projection fibres in the thalamus. Drawings are arranged in the presumptive order of development from A) to E). F) the terminals depicted in B) to the left and those depicted in E) to the right; G) the terminals depicted in B), magnified in focusing on the terminals at the extreme left side. All illustrations were taken from thalamic sections of 6 PO 19. Bars = 50 μm

two distinct types of responses in the frontal motor (areas 4 and 6) and pariental association cortices (areas 5 and 7) respectively. The response in the frontal motor cortex is a diphasic positive-negative wave in the superficial cortical layers and a diphasic negative-positive wave in the deep cortical layers. The response in the parietal association cortex is a negative wave in the superficial cortical layers and a positive wave in the deep cortical layers. BCX transection abolished these responses completely. It is natural because all fibres arising from the lateral and interpositus nuclei cross completely at the BCX. Occurrence of the cerebellocerebral response was tested in 43 kittens and 19 adult cats which were allowed to survive more than 15 days after BCX transection. The cerebellocerebral response could not be evoked in any of the animals in which labelled terminals were virtually absent in the thalamus at the subsequent histological examination. By contrast, marked cerebellocerebral responses were evoked in all of the four kittens (3 PO 34, 6 PO 19, 14 PO 33, and 61 PO 94) and three adult cats in which labelled terminals arising from regenerated axons were distributed densely in the thalamic VA-VL nuclear complex. Marked responses were also evoked in two other kittens (16 PO 31, 32 PO 26) in which the likely-regenerated fibres crossed the midline by coursing around the bottom side of the cutting device (Fig. 6). Responses recorded from the kitten (6 PO 19) on stimulation of the contralateral interpositus nucleus are shown in Fig. 9 A–B for the frontal (A) and parietal cortices (B). Responses recorded from the kitten (14 PO 33) on stimulation of the contralateral lateral nucleus are illustrated in Fig. 9 C–D for the frontal (C) and parietal cortices (D). In four (14 PO 33, 3 PO 34, 16 PO 31, 32 PO 26) of the six kittens and three adult cats, the cerebellocerebral responses were evoked only contralaterally as in intact animals. The responsive area and configuration of response in these animals did not differ from those in intact animals[16]. In the remaining two kittens (6 PO 19, 61 PO 94) in which not only a marked contralateral cerebellofugal projection but also a considerable amount of aberrant ipsilateral projection was observed, the cerebellocerebral responses were evoked bilaterally although the magnitude of the ipsilateral response was small. Again, the responsive area and configuration of response were normal, provided the bilaterality of the response was disregarded. Similar bilateral cerebellocerebral responses have been reported to occur in kittens after neonatal hemicerebellectomy which induces collateral sprouting from intact cerebellothalamic neurons in the spared hemicerebellum[17, 18].

What Are the Factors Inducing Successful or Abortive Regeneration?

Marked, functionally active, regeneration of the cerebellofugal projection occurred unambiguously in kittens as well as in adult cats after transection of the BCX, but why does marked regeneration occur so infrequently as compared to abortive regeneration? What are the factors which bring about successful

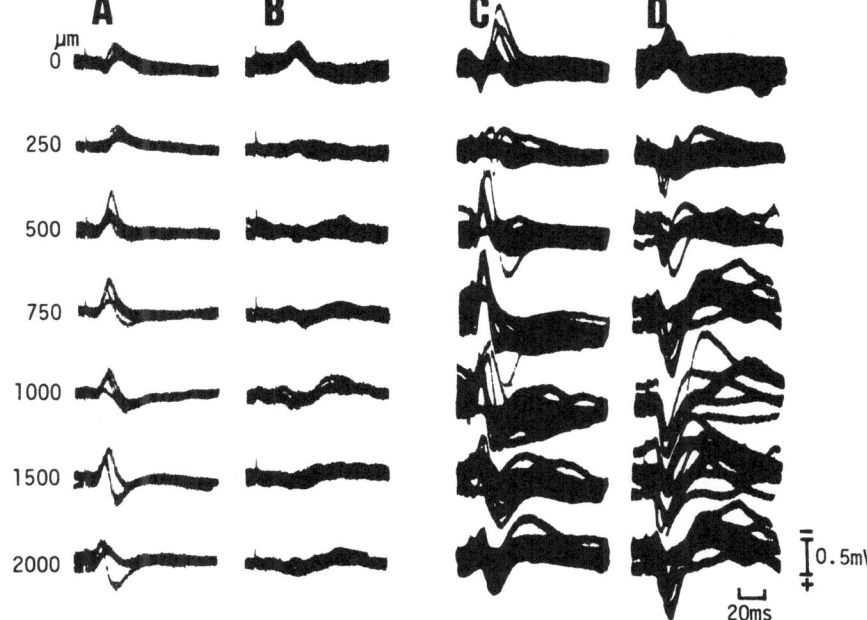

Fig. 9. Cerebellocerebral responses mediated by the regenerated cerebello-thalamic projection. Responses were recorded from the frontal motor (area 4; A and C) and parietal association cortices (area 7; B and D) on stimulation of the interpositus nucleus (A and B) in a kitten (6 PO 19) and on stimulation of the lateral nucleus (C and D) in another kitten (14 PO 33). Morphological findings in the former animal are shown in Figs. 3, 5, and 8; those of the latter animal in Fig. 4. Numerals to the left of the records indicate depths from the cortical surface. Calibrations are applicable to all records

regeneration or abortive regeneration? At the moment we have no definite answer for these questions. When marked regeneration occurred, the initially divided tissue was continuous and dense glial scars were absent, at least in the area of fibre crossing even though proliferation of glia cells occurred diffusely. By contrast, when regeneration became abortive, the lesion had occasionally a cavity and dense glial scars were always present over the entire extent of the lesion irrespective of the presence or absence of a cavity. Such dense glial scars, as an impenetrable barrier, appeared to prevent growing of axons. Therefore, we expected that suppression of dense gliosis might be beneficial for growing of axons. Local administration of arabinosyl-cytosine and 5-fluorodeoxyuridine suppressed formation of glial scars, but however, it did not facilitate axonal regeneration[14]. These chemicals did not suppress the budding of sprouts from transected axons but the sprouts could not cross the gap that was produced by transection and remained unfilled. The persistence of the gap is due to suppression of connective tissue formation by the chemicals. When the gap was bridged by a tungsten or stainless steel wire, numerous regenerated axons occasionally crossed the lesion along the wire[14]. Thus, a lack of dense glial scars does not appear to be a sufficient condition, though it may be a necessary condition, for the sprouts to grow; the sprouts appear to demand a continuous route for their growing. The age of the animal at the time of operation appears unimportant because there was no significant difference between kittens and adult cats for the incidence of the occurrence of marked regeneration. Whether axons are myelinated or unmyelinated appears also unimportant because cerebellothalamic projections are mostly myelinated except during the neonatal period as indicated by the latency of the cerebello-cerebral response[16].

Ramón y Cajal reached the conclusion of abortive regeneration of mammalian central axons, but nevertheless knew their latent regrowth potential and wrote in his book[4] "It is for the science of the future to change, if possible, this harsh decree. Inspired with high ideals, it must work to impede or moderate the gradual decay of the neurones, to overcome the almost invincible rigidity of their connections, and to re-establish normal nerve paths, when disease has severed centres that were intimately associated." We, now, having convincing evidence for the occurrence of successful regeneration, can say that years of pessimism about the failure of mammalian central axons to regenerate are giving way to new optimism about their regrowth potential[19].

Repairing of neural tracts after brain injury may not be an insubstantial dream.

References

1. Aguayo AJ, Benfey M, David S (1983) A potential for axonal regeneration in neurons of the adult mammalian nervous system. Birth Defects 19: 327–340
2. Björklund A, Stenevi U (1979) Regeneration of monoaminergic and cholinergic neurons in the mammalian central nervous system. Physiol Rev 59: 62–100
3. Björklund A, Stenevi U (1984) Intracerebral neural implants: Neuronal replacement and reconstitution of damaged circuitries. Ann Rev Neurosci 7: 279–308
4. Cajal S, Ramón y (1959) Degeneration and regeneration in the nervous system (1928), May RM (transl). Hafner, New York, 1959
5. Clemente CD (1964) Regeneration in the vertebrate central nervous system. Int Rev Biol 6: 257–301
6. Foerster AP (1982) Spontaneous regeneration of cut axons in adult rat brain. J Comp Neurol 210: 335–356
7. Gage FH, Dunnett SB, Stenevi U et al (1983) Aged rats: recovery of motor impairments by intrastriatal nigral grafts. Science 221: 966–969
8. Guth L (1975) History of central nervous system regeneration research. Exp Neurol 48 (Part 2): 3–15
9. Kalil K, Reh T (1979) Regrowth of severed axons in the neonatal central nervous system: establishment of normal connections. Science 205: 1158–1160
10. Kalil K, Reh T (1982) A light and electron microscopic study of regrowing pyramidal tract fibers. J Comp Neurol 211: 265–275
11. Kao CC, Chang LW, Bloodworth JMB (1977) Axonal regeneration across transected mammalian spinal cords: An electron microscopic study of delayed microsurgical nerve grafting. Exp Neurol 54: 591–615
12. Kawaguchi S, Miyata H, Kawamura M et al (1981) Morphological and electrophysiological evidence for axonal regeneration of axotomized cerebellothalamic neurons in kittens. Neurosci Lett 25: 13–18
13. Kawaguchi S, Miyata H, Kato N (1982) Axonal regeneration of axotomized cerebellothalamic projection neurons in adult cats. J Physiol Soc Japan 44: 383
14. Kawaguchi S, Miyata H, Kato N (1984) Mechanical guidance for axonal regeneration of cerebellothalamic neurons in the cat. Neurosci Lett Suppl 17: S 20
15. Kawaguchi S, Miyata H, Kato N (1986) Regeneration of the cerebellofugal projection after transection of the superior cerebellar peduncle in kittens: morphological and electrophysiological studies. J Comp Neurol 245: 258–273
16. Kawaguchi S, Samejima A, Yamamoto T (1983) Post-natal development of the cerebello-cerebral projection in kittens. J Physiol (Lond) 343: 215–232
17. Kawaguchi S, Yamamoto, Samejima A (1979) Electrophysiological evidence for axonal sprouting of cerebellothalamic neurons in kittens after neonatal hemicerebellectomy. Exp Brain Res 36: 21–39
18. Kawaguchi S, Yamamoto, T., Samejima A et al (1979) Morphological evidence for axonal sprouting of cerebellothalamic neurons in kittens after neonatal hemicerebellectomy. Exp Brain Res 35: 511–518

19. Marx JL (1980) Regeneration in the central nervous system. Science 209: 378–380
20. Puchala E, Windle WF (1977) The possibility of structural and functional restitution after spinal cord injury. A review. Exp Neurol 55: 1–42
21. Richardson PM, McGuiness UM, Aguayo AJ (1980) Axons from CNS neurons regenerate into PNS grafts. Nature 284: 264–265
22. So K-F, Aguayo AJ (1985) Lengthy growth of cut axons from ganglion cells after peripheral nerve transplantation into the retina of adult rats. Brain Res 328: 349–354
23. Tsukahara N (1981) Synaptic plasticity in the mammalian central nervous system. Ann Rev Neurosci 4: 351–379
24. Windle WF (1956) Regeneration of axons in the vertebral central nervous system. Physiol Rev 36: 427–440

Abbreviations

3 N Oculomotor nerve; BCX decussation of brachium conjunctivum; Caj interstitial nucleus of Cajal; CL central lateral nucleus of thalamus; CM cetromedian nucleus of thalamus; CN cerebellar nuclei; Da nucleus of Darkschewitsch; FF Forel's field; FTC central tegmental field of the brain stem; INT interpositus nucleus of cerebellum; IO inferior olivary nucleus; Ip interpeduncular nucleus; LAT lateral nucleus of cerebellum; LP lateral posterior nucleus of thalamus; M.E. microelectrode; NCM central medial nucleus of thalamus; NPC nucleus of posterior commissure; PAG peri-aqueductal grey; Pc paracentral nucleus of thalamus; PN pontine nucleus; PRT pretectum; PTN pontine tegmental nucleus; Pul pulvinar nucleus of thalamus; RN red nucleus; SC superior colliculus; S.E. stimulating electrode; VA ventral anterior nucleus of thalamus; VL ventral lateral nucleus of thalamus; VPL ventral posterolateral nucleus of thalamus; ZI zona incerta.

Correspondence: Dr. S. Kawaguchi, Department of Physiology, Institute for Brain Research, Faculty of Medicine, Kyoto University, Kyoto 606, Japan.

Acta Neurochirurgica, Suppl. 41, 18–28 (1987)

Plasticity of Cortical Function Related to Voluntary Movement Motor Learning and Compensation Following Brain Dysfunction

K. Sasaki and **H. Gemba**

Institute for Brain Research, Faculty of Medicine, Kyoto University, Kyoto, Japan

Summary

Processes of motor learning and of compensation after localized brain dysfunction were studied in the monkey tasked with conditioned (visually-initiated, reaction-time) hand movements. Field potentials in various cortical areas of the cerebral hemisphere were recorded successively for many months with electrodes implanted on the surface and in the depth of the cortex. The potentials associated with the conditioned movement were found to change during processes of learning the movement and during courses of degradation as well as recovery after the brain dysfunction.

Motor learning of the reaction-time movement can be categorized into "recognition" and "skill" learnings. The former is the process for associating the visual stimulus with the movement and accompanied mainly by increasing activities of prefrontal, premotor and prestriate cortices. The latter is attaining better performances in executing the movement, particularly with shorter and more fixed reaction time, and is accompanied by recruitment of the cerebro-cerebellar interaction.

The conditioned movements which had been well established were experimentally disturbed by transient, local cooling of different cortical areas or by cerebellar hemispherectomy. Three possible mechanisms of compensation are proposed as follow:

1. *Substitution*: Compensation occurring immediately through substitutional neuronal circuits. On transient cooling of the forelimb motor cortex, the somatosensory cortex became predominant in motor function and replaced the disabled motor cortex in executing the reaction-time movement, although activity was weak and slow (paretic but not paralytic). Resection of the cerebellar hemisphere also induced the compensatory motor function of the somatosensory cortex so that the movement could be performed although it was weak, slow and clumsy.

2. *Relearning*: Compensation by relearning through normally unused neuronal circuits. Prolonged and variable reaction times after cerebellar hemispherectomy persisted when the operation included both dentate and interpositus nuclei but recovered within about three weeks when the interpositus nucleus was preserved. It is suggested that the information processing for the well accomplished reaction-time movement is mainly mediated by the cerebro-cerebellar neuronal circuit including the dentate nucleus but is gradually relearned through the normally unused circuits involving the interpositus nucleus after the dentate nucleus lesion.

3. *Rebuilding*: Compensation by rebuilt neuronal circuits, *e.g.*, by sprouting and/or regeneration (see S. Kawaguchi in this book).

Keywords: Motor learning; compensation; cortical field potential; cerebro-cerebellar interaction; conditioned hand movement; monkey.

Introduction

Spontaneous recovery of function after the brain has been damaged is one of the most remarkable features of brain mechanisms, though the extent of recovery varies depending upon the site and severity of the lesion. The processes of the recovery after brain damage often appear to be similar to those of the learning in an untaught normal subject that is, for instance, attaining a purposeful movement, *i.e.*, voluntary movement. In this paper, we will briefly review some of our experimental studies on learning processes of conditioned hand movements and also on compensatory processes for the movement disturbed by localized dysfunction of the cerebral cortex or the cerebellum.

Methods

Recording Cortical Field Potentials in Monkeys Performing Conditioned Hand Movements

The data shown in this paper were obtained in our experimental series where we used more than 30 adult monkeys (Macaca fuscata). They were trained to lift a lever with wrist extension in response to a light stimulus. The stimulus lasted for about 500 ms and was delivered at random time intervals of 2.5–6.0 s. The monkey had to move the hand within the duration of the visual stimulus given in front of the monkey by a diode emitting green light (V. S. in Fig. 1 D). At first the monkey moved the hand in his own time without regard to the light stimulus but gradually learned to perform the reaction-time movement (see Fig. 4).

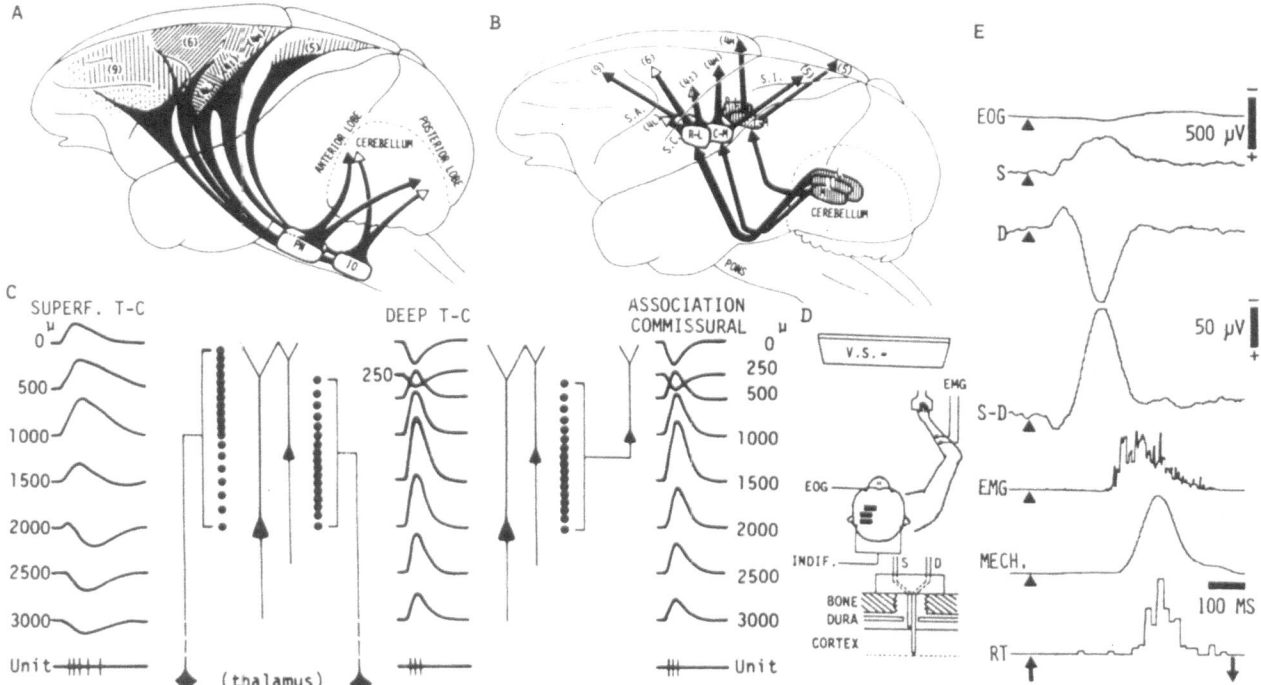

Fig. 1. A) and B) Diagram of cerebro-cerebellar interconnections in monkeys. Numbers indicate the different cortical areas after Brodmann. Area 4 is divided into lateral (*4 L*), intermediate (*4 I*) and medial (*4 M*) parts. *PN* pontine nuclei. *IO* inferior olive. *S.A.* arcuate sulcus. *S.C.* central sulcus. *S.I.* intraparietal sulcus. *M, I, L* medial, interpositus, lateral nuclei of the cerebellum. *R-L, C-M* two nuclear complexes of the thalamus. C) Diagram of laminar field potentials in the cortex by the two thalamo-cortical (superficial and deep T-C) and the cortico-cortical (association and commissural) inputs. Numbers indicate the depth from the cortical surface in μm. Unit, schematic pattern of firing of cortical pyramidal neurone. The presumed excitatory synaptic terminals on apical dendrites of cortical pyramidal neurones in layer III and V are diagramatically shown by dots for the three afferent inputs. Laminar field potentials are attributed to electrical dipoles generated in pyramidal neurones by the EPSPs. D) Diagram of the methods of chronic experiments with monkeys, as described in the text. In the lower part, electrodes placed on the cortical surface (*S*) and in 2.0–3.0 mm depth (*D*) through the bone and dura are illustrated. E) Examples of EOG, cortical potentials (*S, D, S-D*), *EMG*, mechanogram of lever movement (*MECH.*) and reaction time histogram (*RT*) are presented as explained in the text. Calibration of 500 μV for EOG and 50 μV for cortical potentials. 100 ms time scale for all traces. (Modified from Sasaki, 1979)

Electrodes for recording cortical field potentials consist of silver needles (about 200 μm in diameter) insulated except for their pointed tips. A pair of them were placed respectively on the surface and at 2.0–3.0 mm depth in the cerebral cortex and fixed to the bone by dental cement (Fig. 1 D, lower part). Usually 10–20 pairs of such electrodes were implanted in each hemisphere and served for recording potentials for several months. Cortical field potentials were recorded against indifferent electrodes buried in the bone behind the ear on both sides (INDIF. in Fig. 1 D), amplified through 2.0 s time constant amplifiers and recorded on magnetic tapes, together with pulses of stimulus onset and movement onset etc. Electro-oculogram (EOG) led from the rostrolateral edge of the frontal bone was monitored (EOG in Fig. 1 D). Examples are shown in Fig. 1 E of EOG, surface (S), depth (D), surface minus depth (S—D) potentials from the forelimb motor cortex contralateral to the hand being tested, being averaged 100 times by the stimulus onset pulse (triangle and upward arrow). Electromyogram (EMG) recorded bipolarly from the skin over wrist extensor muscles was rectified and likewise averaged, and the lever movement was converted to electrical signal and aligned 100 times (MECH.). Reaction times of the 100 movements are given in a histogram of 16 ms bins (RT) where the onset and end of the stimulus are indicated by upward and downward arrows respectively.

Cortical field potentials are interpreted as a temporal and spatial summation of responses in the cortex induced by thalamo-cortical (T-C) and/or cortico-cortical (C-C) volleys which have been analysed by microelectrophysiological experiments in the monkey (Fig. 1 C) (see Sasaki, 1979). For instance, potentials in the forelimb motor cortex (Fig. 1 E) can be explained as follow: A visual volley arrives in the cortex at a latency of about 50 ms mainly through deep T-C projections (surface-positive, depth-negative; s-P, d-N potentials) and also induces cerebello-thalamo-cortical impulses which activate the forelimb motor cortex via superficial T-C projections at a latency of 100–150 ms (s-N, d-P potentials) to execute the hand movement (see Sasaki and Gemba, 1982). The neocerebellum that sends volleys to the forelimb motor cortex is activated presumably by the prefrontal and premotor cortices according to the previous studies on cerebro-cerebellar interconnection (Fig. 1 A and B) (see Sasaki, 1979). Therefore it may be assumed that the motor command triggered by the visual stimulus is mediated by the pathway as follow: prefrontal and premotor cortex—pontine nuclei and inferior olive—cerebellum—thalamus—motorcortex—motoneurone. This assumption has been supported by experimental investigations as partly shown in the present paper. Besides the motor cortex, various areas of the cerebral hemispheres on both sides work actively in association

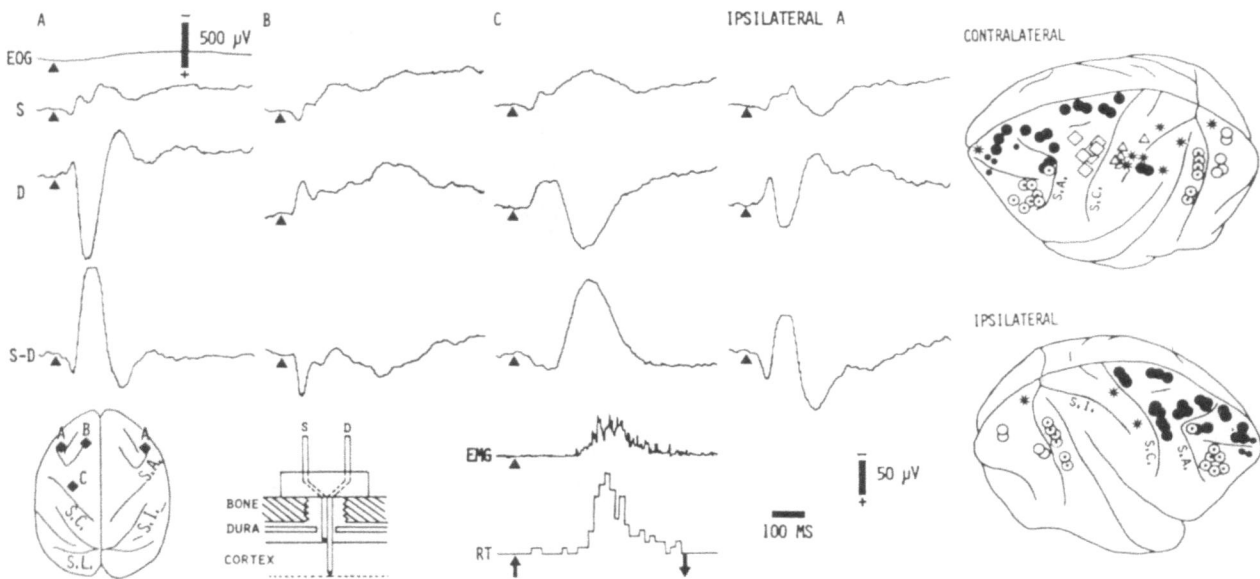

Fig. 2. Cortical field potentials associated with visually-initiated hand movements and their distribution in both cerebral hemispheres. As shown in the lower left diagrams, surface (*S*) and depth (*D*) electrodes in four cortical loci (A-C, IPSILATERAL A) revealed potentials in *S, D* and *S-D* rows respectively. All cortical potentials and EOGs were aligned 100 times by the onset pulse of stimulus (triangle and upward arrow). Other abbreviations and explanations are same as in Fig. 1 and described in the text. Calibration of 500 µV for EOG and 50 µV for cortical potentials. 100 ms time scale for all traces. The recording sites are summarized from 14 monkeys at the well trained stage and plotted on the dorsolateral aspects of the hemisphere contralateral and ipsilateral to the moving hand on the right. Different symbols represent different kinds of potentials. Loci without marked potentials are indicated by asterisks. Some of the electrode sites could not be plotted since they were too crowded in a few areas. (Modified from Sasaki and Gemba, 1982)

with the visually-initiated, reaction-time hand movement as shown in Fig. 2. Particularly remarkable potentials can be noted in the prearcuate area of the prefrontal cortex, premotor, somatosensory, prestriate area etc. in addition to the forelimb motor area.

Local Cooling of the Cortex

The contribution of each cortical area to the conditioned movement and functional significance of cortical potentials on the movement were investigated by cooling temporarily each area and by observing changes of the movement and cortical potentials thereby. For local cooling, metal chambers of which the undersides were shaped so as to cover respective cortical areas to be cooled were placed on several areas in each hemisphere and fixed to the bone by dental cement (Fig. 3). Every chamber had input and output pipes which were connected to a pump for perfusing warm (38–39 °C) or cold (about 1 °C) water. The silver needle electrodes were set under the chamber to record cortical field potentials in the control and cooling states. At first, control records were obtained under all chambers perfused with warm water. Then only the chamber to be cooled was switched to cold water, and cortical potentials and movements were observed during the cooling. After cooling, the chamber was perfused again with warm water and recovered states after rewarming were recorded. Time course of cooling and temperature gradient in the brain tissue under cooling and warming chambers are shown in Fig. 3 and 8 (see Sasaki and Gemba, 1984a).

Cerebellar Hemispherectomy and Histological Examination

After the monkey had been well trained with the reaction-time hand movement. the cerebellar hemisphere ipsilateral to the operant hand was resected to observe changes of the movement and cortical field potentials due to the operation. In some monkeys, cerebellar hemispherectomy was done before teaching the movement in order to study the cerebellar contribution to motor learning. The hemispherectomy usually included both the dentate and interpositus nuclei but occasionally only the dentate nucleus as will be described later. The extent of the operation was checked histologically after electrophysiological examinations had lasted for several months (Sasaki et al., 1982). The sites of recording electrodes in the cortex were also examined morphologically after the experiments.

Results and Discussion

Development and Change of Cortical Field Potentials During Learning Processes of Conditioned Hand Movement

Field potentials in various cortical areas were successively measured for several months in which untaught monkeys gradually learned the reaction-time hand movement in response to the light stimulus and reached the well established state. Fig. 4 presents examples of seven cortical loci (A-G) contralateral to the moving hand in one and the same monkey. As seen in the reaction time (RT) trace, the monkey initially lifted the lever in its own time without regard to the light stimulus (900 ms duration starting at the upward

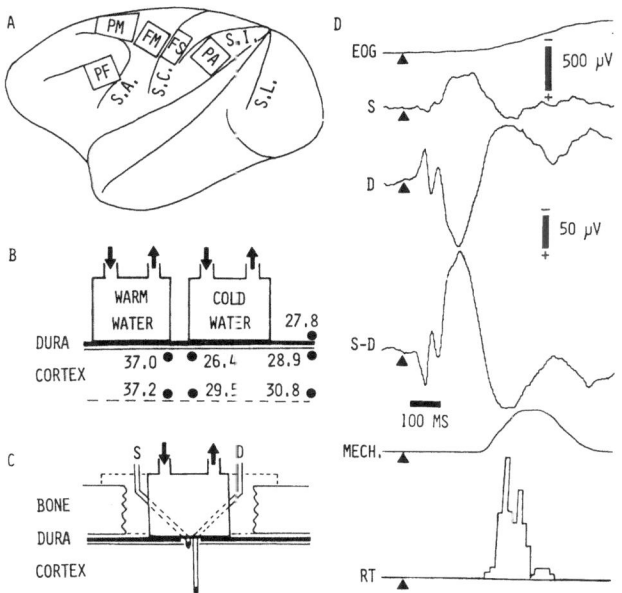

Fig. 3. Arrangement of local cooling of the cerebral cortex and recording of premovement cortical field potentials. A) The underside shape of cooling and warming chamber is drawn on the lateral view of the left hemisphere. *PF* a part of the prefrontal cortex. *PM* premotor cortex. *FM* forelimb area of the motor cortex. *FS* forelimb area of the somatosensory cortex. *PA* a part of the parietal association cortex. *S.L.* lunate sulcus. Other abbreviations are same as in Fig. 1. B) Two chambers for warming and cooling placed on the dura and temperature (in °C) in the cortex which was measured in the steady state after 10-minute circulation of warm (38–39 °C) and cold (about 1 °C) water. C) Recording electrodes placed on the surface (*S*) and at 2.0–3.0 mm depth (*D*) of the cortex under a chamber. The chamber and the electrodes were fixed to the bone by dental cement (broken lines). D) Examples of EOG and field potentials in the forelimb motor cortex as in Fig. 1 E. (Sasaki and Gemba, 1984 a)

arrow, the end part being curtailed). Only the movements which occurred incidentally during the light stimulus were sampled in the RT histogram. Alignment of cortical field potentials (surface minus depth potentials in A–G rows) by the onset pulse of stimulus revealed some significant responses in the prefrontal (A–C), premotor (D), prestrite (F) and striate (G) cortices even at this untaught stage (column I). These responses gradually increased in size on each day of

training (column II) until the stage III when the monkey started to move the hand in response to the light stimulus (column III). Reaction times were longer and more variable at stage III than stage IV (column IV) and many trials occurred out of the stimulus duration without obtaining reward. At this stage, surface-positive, depth-negative (s-P, d-N) potentials appeared in the forelimb motor cortex (E row) at a latency of about 50 ms.

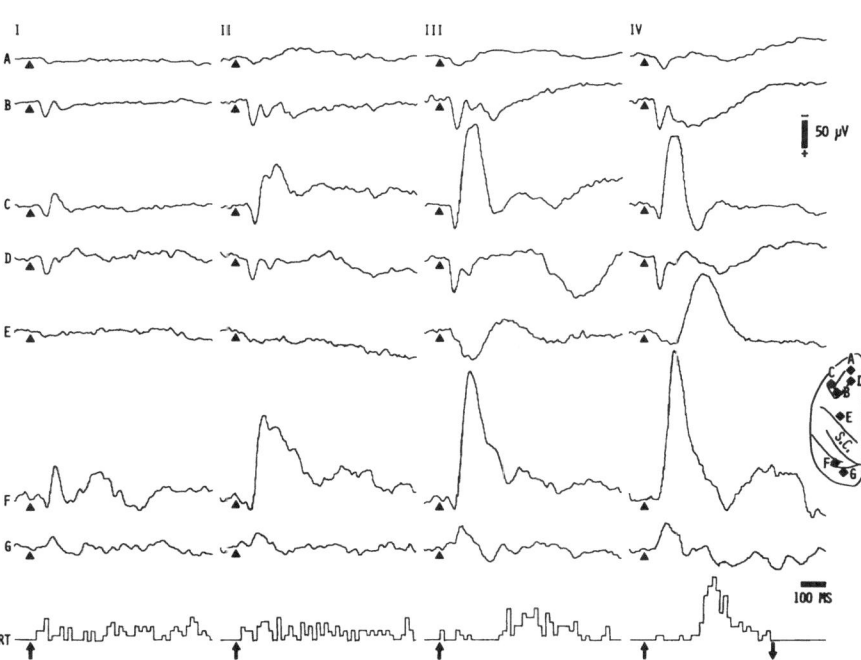

Fig. 4. Premovement S-D potentials in seven cortical loci on the left hemisphere of a monkey learning the reaction-time movement of the right hand. Columns *I–IV* present four sessions at different stages. *I* was taken in the second session, *II* 21 days after *I*, *III* 3 days after *II*, and *IV* 36 days after *III*. Potentials were averaged 100 times from every session for respective cortical loci indicated by alphabetical symbols in the inset diagram and each row of records. The stimulus was given for 900 ms at stage *I–III* but 510 ms at *IV*. Only the movements which occurred during the light stimulus were counted in RT histograms and averaged potentials (the later part of 900 ms is curtailed in columns *I–III*). They were aligned by the stimulus onset pulse as indicated by triangle and upward arrow, the downward arrow showing the end of 510 ms stimulus. Calibration of 50 μV for all potentials and 100 ms for all traces. (Sasaki and Gemba, 1982)

After stage III, repeated trainings hardly changed the potentials in association and premotor cortices but added gradually the s-N, d-P potential in the forelimb motor cortex at a latency of about 120 ms in parallel with faster and less variable reaction times, *i.e.*, better timing of the movement (column IV, E and RT rows). The s-N, d-P potential in the forelimb motor cortex can be attributed to the response elicited by the volley through the neocerebellum and the thalamus, which has been supported by several sorts of experiments as partly exemplified later (see Sasaki et al., 1982). Thus it may be considered that the gradual recruitment of cerebro-cerebellar interaction is closely correlated with attaining fast and fixed reaction-time movements. Quick and appropriate timing is undoubtedly one of the most important factors of skilled intentional movements, and the process from stage III to IV may be called "skill learning" as in contrast to the process from stage I to III which can be called "recognition learning" in associating the visual stimulus with the movement (Sasaki and Gemba, 1982).

The process from stage III to IV is demonstrated more in detail in Fig. 5. After stage III (the left column), records at one week interval are shown successively to the right, being from the premotor (A row) and forelimb motor (B rows) cortex respectively. The same data from the motor cortex are aligned 100 times not only by the onset pulse of stimulus (upper B row in V.S.

part) but also by that of movement (lower B row in L.E. part). It is clearly noted that reaction times were gradually improved in accordance with increase of the s-N, d-P potential in the motor cortex. The heights of the potential (μV) and the reaction time (ms) at the peak of histogram are plotted on ordinates against days after the beginning of stage III on abscissae in Fig. 6 for two monkeys which required the longest (A) and the shortest (B) times from stage III to IV among fifteen monkeys measured. Close correspondence can be seen between shortening of reaction times and increasing of s-N, d-P potentials in both cases, revealing the contribution of cerebro-cerebellar interaction to the skill learning.

The idea that the recruitment of cerebro-cerebellar interaction is indispensable to skill learning was supported also by the experiment in which the cerebellar hemisphere was ablated before training the monkey for the reaction-time movement as shown in Fig. 7. After the cerebellar hemispherectomy, the monkey was taught the movement and found to be able to progress in the recognition learning but not in the skill learning as seen in the records from column I to III and from III to IV respectively. In fact, in several weeks the monkey was repeatedly trained without appreciable improvement in the reaction times (from III to IV), although the monkey had associated the visual stimulus with the reaction-time movement (from I to III). Certainly no

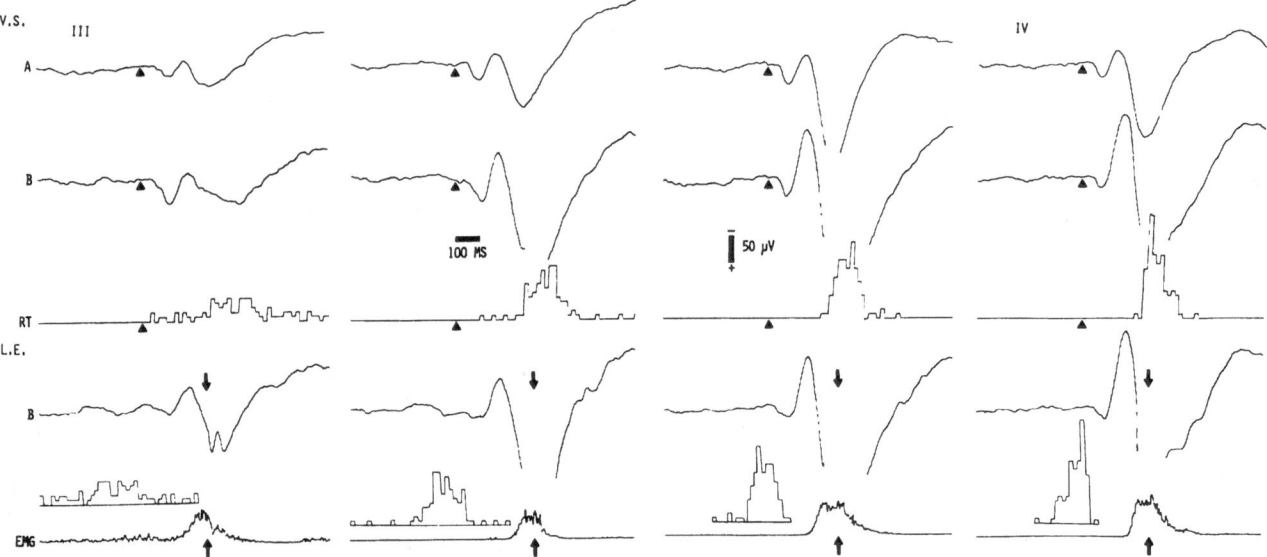

Fig. 5. Premovement S-D potentials in the premotor (*A*) and forelimb motor (*B*) cortices are presented in four different sessions at one week interval from stage *III* (leftmost column) to stage *IV* (rightmost column). V.S.: Two rows of the potentials (*A* and *B*) are aligned 100 times by the stimulus onset (triangle) with RT histograms. *L.E.*: The same data as the row B in V.S. are aligned by the movement onset pulse (arrows). Histograms of the onset time of light stimulus preceding the movement are shown above the EMGs aligned to the onset of movement. The light stimulus lasted for 900 ms in the leftmost and second columns and 510 ms in the third and rightmost columns. Calibration of 50 μV for all cortical potentials and 100 ms for all traces. (Sasaki and Gemba, 1983)

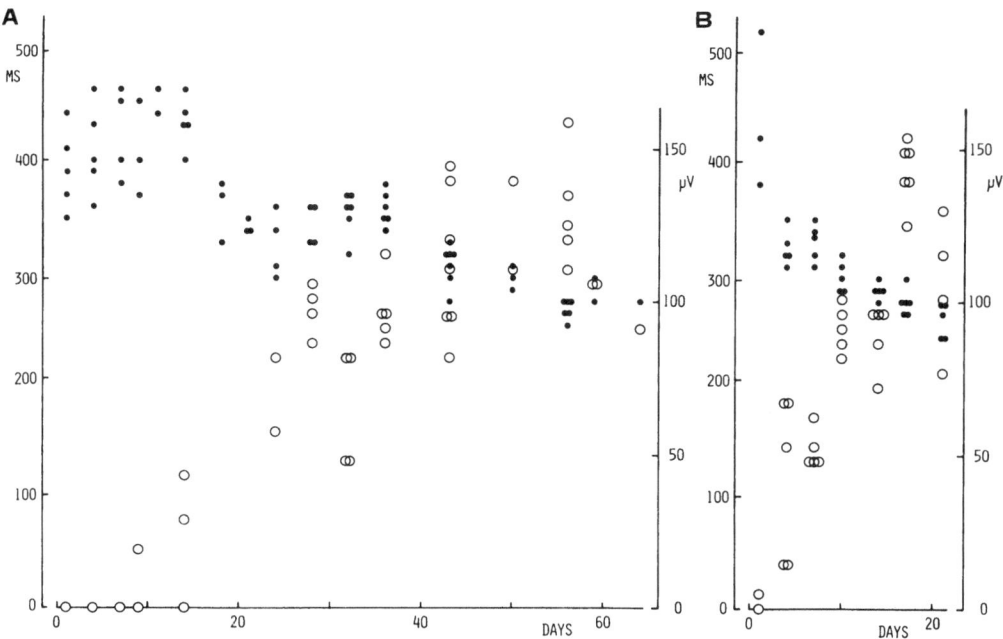

Fig. 6. Reaction times in ms (filled circle) and heights of s-N, d-P field potentials in the forelimb motor cortex in μV (open circle) are plotted on ordinates, against days from stage III to IV on abscissae. Examples of two different monkeys are presented in A) and B) respectively. The reaction time represents that of the peak in the histogram of 100 samples in every session. For the reaction time and the potential height, several points are given in respective sessions on each day. (Sasaki and Gemba, 1983)

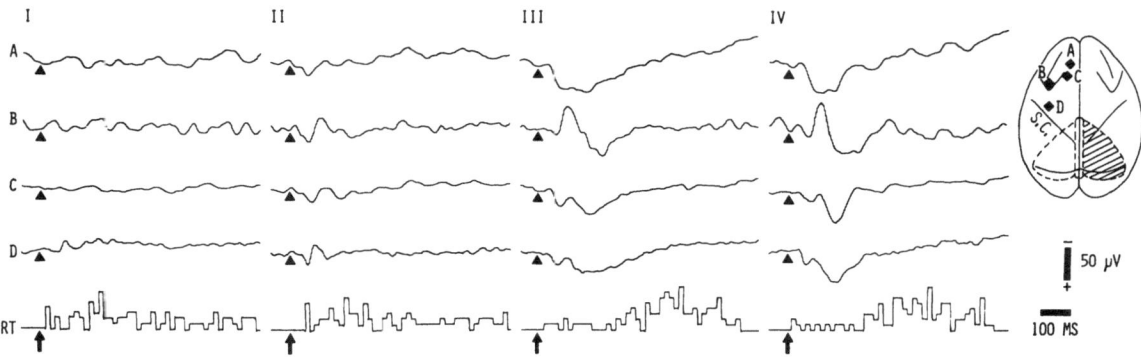

Fig. 7. Premovement S-D potentials in four loci of the left cerebral hemisphere of a monkey whose cerebellar hemisphere on the right side (shaded by oblique lines) had been extirpated before training of the reaction-time movement of the right hand. The potentials were aligned 100 times by the onset pulse of light stimulus lasting for 900 ms. Their later part is curtailed. Calibration of 50 μV for all potentials and 100 ms for all traces. (Sasaki and Gemba, 1982)

s-N, d-P potentials were seen in the forelimb motor cortex (D row) contralateral to the resected cerebellar hemisphere and the moving hand. It should be noted that the recognition learning can be achieved without the neocerebellum.

Compensation After Localized Cooling or Ablation of the Brain

After the monkey learned the visually-initiated, reaction-time hand movement, a localized area of the cerebral cortex was cooled or the cerebellar hemisphere was ablated to study the functional significance of the cortical area and the cerebellar hemisphere (neocerebellum) in the movement. References so far reported (see S. Kawaguchi in this book) and our own studies suggest that there are considered to be three compensatory mechanisms which follow localized lesion in the brain. (1) *Substitution*: Compensation occurring immediately through substitutional neuronal circuits. (2) *Relearning*: Compensation by relearning through normally unused circuits. (3) *Rebuilding*: Compensation by rebuilt neuronal circuits (sprouting and regeneration). Experimental findings suggesting the substitution and the relearning will be presented below.

Temporary local cooling of the forelimb motor cortex.

As shown in Fig. 2, various cortical areas of both cerebral hemispheres show activities in association with the visually-initiated hand movement. In order to study the contribution of each area to the movement and to learn the functional significance of cortical potentials in the respective areas, local cooling of different areas was tried in the monkey performing the movement (Fig. 3) (see Sasaki and Gemba, 1984 a, b; 1986). The cooling method briefly described earlier was first applied to the forelimb motor cortex, since we expected simply that the cooling might have induced paralysis of contralateral forelimb muscles and therefore be suitable for testing the cooling device.

The time course of the motor cortex cooling is illustrated in Fig. 8. Measurement of temperature in the cortex under the chamber was made after all electrophysiological experiments had been finished and revealed data shown in Figs. 3 B and 8 A. Warming chambers surrounding a cooling chamber were quite effective in localizing the cooling effect to the cortical region under the cooling chamber (Fig. 3 B) (see Sasaki and Gemba, 1984 a). Temperature under the cooling chamber went down to a steady level between 20–30 °C within five minutes and was raised up to the control level within several minutes by rewarming (Fig. 8 A). Field potentials under the cooling chamber were depressed immediately after the start of cooling (Fig. 8 B), the force of wrist muscles was weakened and reaction time was prolonged with much the same time course as field potentials (Fig. 8 C and D). These changes were reversible by rewarming. The results indicate that cooling of the forelimb motor cortex results in paresis of forelimb muscles instead of paralysis.

Simultaneous recording from the somatosensory cortex disclosed that the forelimb somatosensory cortex increased its premovement activity and became predominant in motor function during dysfunction of the motor cortex as shown in Fig. 9. Field potentials (surface minus depth potential, S-D) in the motor cortex preceded the movement more than those in the somatosensory cortex under normal conditions (Fig. 9 A). When cooled, potentials in the motor cortex were decreased considerably but those in the somatosensory cortex increased in size and preceded the movement much more predominantly than in the control state or even more than the motor cortex does

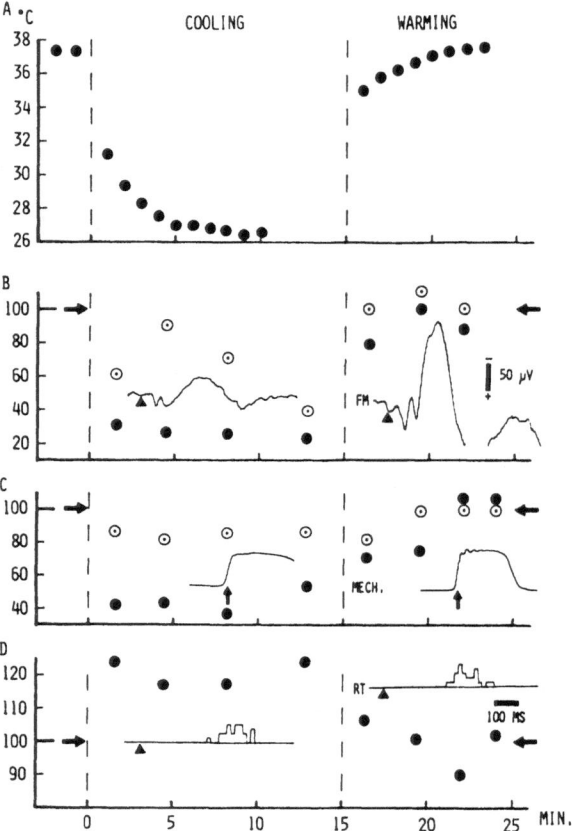

Fig. 8. Changes of cortical temperature (A), premovement cortical potentials in the forelimb motor cortex (B), muscular tension (C) and reaction time (D) during cooling of the cortex in a monkey. A) A thermocouple was placed at about 2 mm depth from the cortical surface under the cooling chamber after the end of experiments shown in B-D. Before and after cooling (between two broken lines), warm water was circulated through the chamber. B) Peak heights of early s-P, d-N premovement potentials are plotted by open circle with dot in centre, and those of late s-N, d-P ones by filled circle. C) Peak height of muscular tension is given by open circle with dot in centre, and maximal rate of rise of the mechanogram by filled circle. D) Mean reaction time. In B-D, percentage of the control value (arrow) is taken in ordinates. Sample records under the control condition (with warming water) and those under the cooling condition are presented respectively in the cooling and warming columns. Cortical potentials in B and reaction time histogram in D were aligned by the stimulus onset pulse (triangle), whereas mechanograms in C were by the movement onset pulse (arrow). 50 µV calibration for cortical potentials in B and 100 ms for all sample records. All abscissae are minutes after the start of cooling. (Modified from Sasaki and Gemba, 1984 a)

under normal conditions (Fig. 9 B). It should be noted that the movement under the motor cortex cooling was certainly delayed (slow) and weak as seen in -RT and MECH./T rows. Yet the data suggested the compensatory motor function of the somatosensory cortex for the disabled motor cortex. This was supported by simultaneous cooling of both motor and somato-

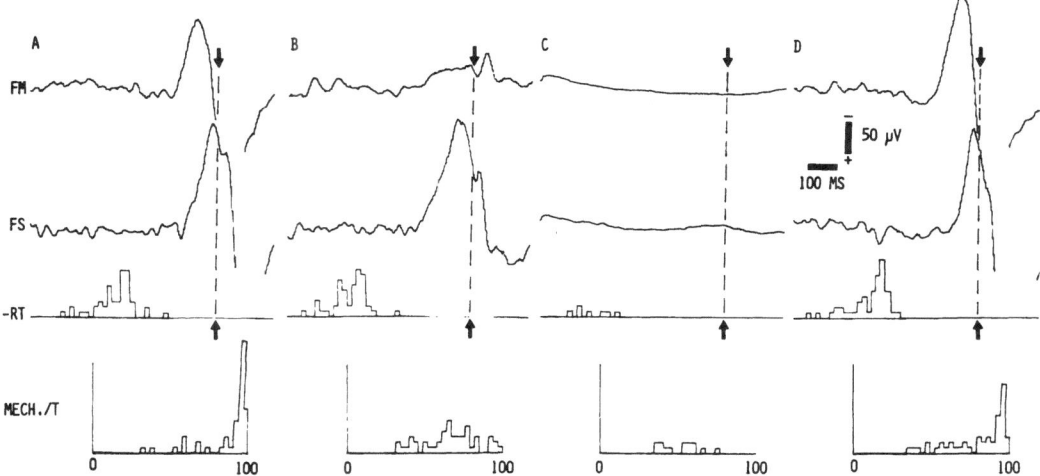

Fig. 9. Paralysis of forelimb muscles due to simultaneous cooling of the forelimb motor (*FM*) and somatosensory (*FS*) cortices in a monkey. A) Control. B) Cooling of the motor cortex. C) Cooling of the motor and somatosensory cortices. D) Rewarming of the cortices. All cortical potentials were averaged by the movement onset pulse (arrows with broken line), 50 times in A, B and D, and 7 times in C. 50 μV calibration is applicable to A, B and D but not to C. The onset time of visual stimulus is plotted in histograms, thus resulting in reversed -RT (in 16 ms bins). Maximal rate of rise in the mechanogram is plotted in histogram for the same 50 trials in percent of the abscissa (3 percent bins) (MECH./T of A, B and D), the fastest among all trials being taken as 100 percent. In C, 14 trials including 7 successful (to obtain reward) and 7 unsuccessful movements are plotted in MECH./T. The cortical potentials and reaction time histogram in C are for the 7 successful movements. 100 ms scale for all cortical potentials and reaction time histograms. (Sasaki and Gemba, 1984 a)

sensory cortices which induced a complete paralysis as illustrated in Fig. 9 C. In fact, the monky could lift the lever only seven times at the beginning of simultaneous cooling and its forelimb contralateral to the cooled motor cortex was found to have an entirely flaccid paralysis. These changes were perfectly reversible and reproducible, and they occurred immediately within a minute or so when cooling or rewarming. This appears to be a good example of *substitution* type of compensation.

Cerebellar hemispherectomy. Cerebellar hemispherectomy was found to be associated with both *substitution* and *relearning* compensation as will be exemplified separately.

(i) *Substitution.* In a monkey well-trained with the visually-initited hand movement, cerebellar hemispherectomy ipsilateral to the hand being tested downgraded the learned movement from stage IV to III, *i.e.*, it made the reaction time longer and more variable than before the operation. Concomitant recording from the motor and somatosensory cortices revealed that the somatosensory cortex replaced the motor cortex for the motor function after the cerebellar hemispherectomy just like the cooling of motor cortex as mentioned above (Sasaki and Gemba, 1984 b). Field potentials (S-D) recorded from the forelimb motor (FM) and somatosensory (FS) cortices before (PREOP.) and after (POSTOP.) cerebel-

lar hemispherectomy are averaged 100 times by the stimulus onset (V.S.) and movement onset (L.E.) pulses respectively in Fig. 10 A. As seen in traces in V.S. averages, the operation decreased the s-N, d-P potential in the motor cortex considerably and made reaction times longer and more variable. The L.E. average demonstrates that the s-N, d-P potential in the somatosensory cortex was enlarged and preceded the movement much more than before the operation. The premovement area above the baseline in the somatosensory potential is plotted on ordinate, against days after the operation on abscissa in B for three monkeys with cerebellar hemispherectomy. All the three cases (three different symbols) show remarkable enhancement (200–300 per cent of the size before operation) at several days after the operation which quickly declines within 30–40 days. This suggests that the somatosensory cortex compensated temporarily for the motor function of the motor cortex that could not receive the motor command through the neocerebellum.

Somatosensory cortex cooling supported this suggestion as shown in Fig. 11. In fact, the somatosensory cortex cooling at five days after the operation stopped the reaction-time hand movement completely and the monkey just held the lever without responding to the light stimulus. The forelimb muscles of the monkey under such conditions did not seem to be paralytic,

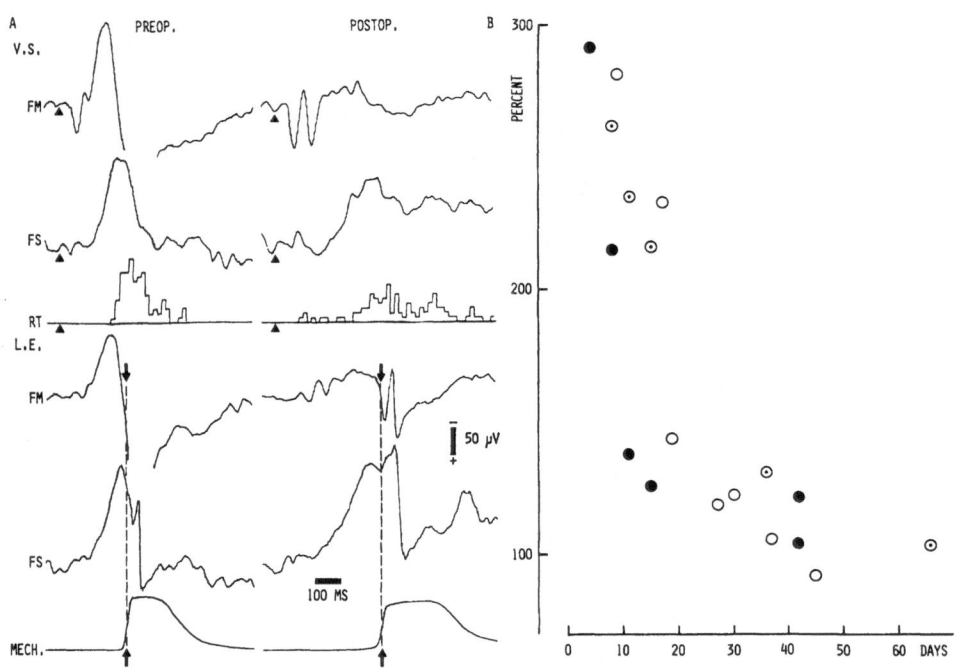

Fig. 10. A) Field potentials (S-D) from the forelimb motor (*FM*) and somatosensory (*FS*) cortices associated with visually-initiated hand movements before (*PREOP.*) and after (*POSTOP.*) cerebellar hemispherectomy in a monkey. The same data were aligned 100 times by the stimulus onset (triangle) (upper half, *V.S.*) and the movement onset (arrows with broken line) (lower half, *L.E.*) pulses respectively. The V.S. averages are accompanied with reaction time histograms (*RT*) and the L.E. averages are with mechanograms (*MECH.*). 50 μV for all cortical potentials and 100 ms for all traces. B) Enhancement of the premovement component of the somatosensory potentials after cerebellar hemispherectomy is plotted against days after the operation (abscissa). The premovement area of S-D potential is plotted on the ordinate as a percentage of the pre-operative value. The three symbols indicate data from three different monkeys. (Modified from Sasaki and Gemba, 1984 b)

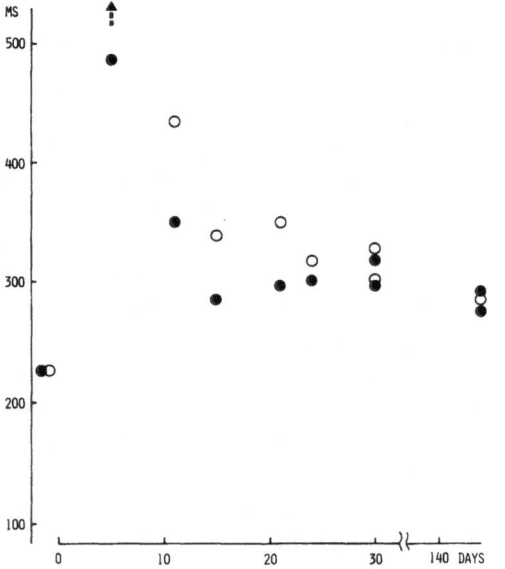

Fig. 11. Effects of the somatosensory cortex cooling upon the visually initiated hand movement after the cerebellar hemispherectomy. The median of 100 reaction times in each session is plotted on ordinate (*MS*) against days after the operation on abscissa respectively with (open circle) and without (filled circle) cooling. Near to the ordinate, the controls before the operation are given. Five days after the operation, the cooling stopped the movement entirely as shown by the arrow (infinite reaction time). (Sasaki and Gemba, 1984 b)

which was different from the simultaneous cooling of both motor and somatosensory cortices mentioned above. The cooling of the somatosensory cortex alone hardly produces any significant disturbance to the reaction-time movement in normal (without cerebellar hemispherectomy) monkey (Sasaki and Gemba, 1984 a). As plotted in Fig. 11, mean reaction time before the operation was not influenced at all by the somato-sensory cortex cooling (leftmost filled and open circles), whereas reaction time became infinite (arrow) with the cooling, five days after the operation, the cooling effect decreasing gradually within about 30 days. These data appear to imply that when disconnected from the neocerebellar input the motor cortex cannot send the motor output to the lower centre, and that the somato-sensory cortex compensates immediately after the operation for the indirectly dysfunctioned ("decerebel-lated") motor cortex. Such a compensatory motor function of the somatosensory cortex appears to be replaced by some other compensatory function gradu-ally developing within 30 days or so, its mechanism not yet being known.

(ii) *Relearning.* When cerebellar hemispherectomy

included both dentate and interpositus nuclei, prolonged reaction times persisted for many months, although some of the other cerebellar symptoms improved to a certain extent. However, in the cases in which the operation involved the dentate nucleus but saved the interpositus nucleus, prolonged reaction times return to the pre-operative level in about three weeks as shown in Fig. 12. The recovery of reaction times was closely correlated with the reappearance and increase in size of the s-N, d-P potential in the forelimb motor cortex, which was entirely eliminated immediately after the cerebellar hemispherectomy (Fig. 12). The time courses of recoveries in the potential and in the reaction time appeared respectively to be similar to those of appearance of the potential and of shortening of reaction times in the learning process in a normal untrained monkey as shown previously (Fig. 4–6). The recovery of s-N, d-P potential occurred only in the cases in which the interpositus nucleus was preserved, and the potential started to reappear as early as eight days after the operation.

Previous microelectrophysiological studies on the cerebro-cerebellar interconnection in monkeys demonstrated that the projection area in the forelimb motor cortex from the dentate nucleus is overlapped considerably by that from the interpositus nucleus (Sasaki et al., 1976). This finding together with the results mentioned just above suggests the most reasonable explanation as follows: In a normal situation, the cerebro-cerebellar interaction for initiating the reaction-time movement includes the dentate nucleus, and the interpositus nucleus is mainly involved in some other motor control system, but when the dentate nucleus is removed or inactivated, the interpositus nucleus is utilized to replace the dentate nucleus. The neuronal circuits including the interpositus nucleus would require some time for the skill learning as do those mediated by the dentate nucleus in the normal untaught state (Sasaki et al., 1982). This appears to be an example of *relearning* with normally unused but existing neuronal circuits after a localized brain lesion.

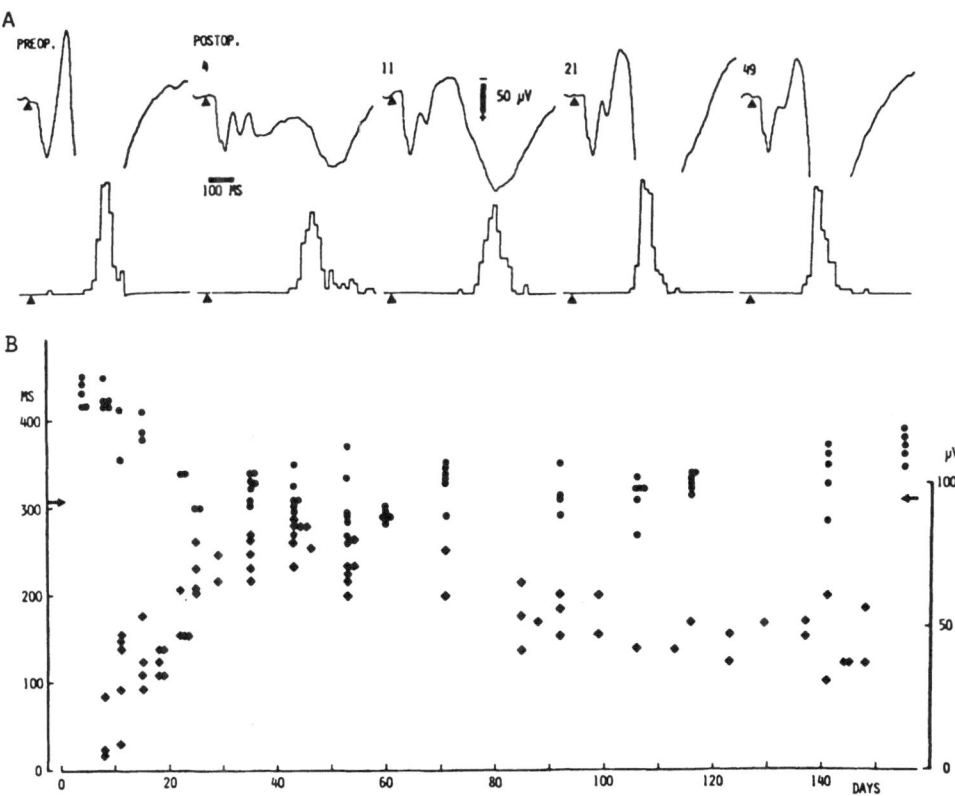

Fig. 12. A) Visually-initiated premovement potentials (S-D) in the forelimb motor cortex (upper row) and histograms of reaction times (lower row) as influenced by cerebellar hemispherectomy when the interpositus nucleus was spared. Numbers indicate days after the operation. 50 μV calibration for potentials and 100 ms scale for all traces. One hundred samples for every potential and histogram. B) Reaction times (ordinate in ms to the left) (filled circle) and the peak heights of the late s-N, d-P potentials (ordinate in μV to the right) (filled square) plotted against days after cerebellar hemispherectomy (abscissa). Mean pre-operative reaction time is indicated by horizontal arrows. (Modified from Sasaki et al., 1982)

Comment

Compensatory functions related mainly to the motor and somatosensory cortices and the cerebellum were taken into account in this paper. These cortices are the primary input (sensory) and output (motor) stations in the cerebral cortex and are considered to be relatively simple in their integrative functions especially when compared with association cortices. In fact, the effects of cooling the premotor cortex, for instance, are more complicated than the motor and somatosensory cortices and dissociate the visuo-motor integration, *i.e.*, degrading the well established conditioned movement to nearly random movement, from stage IV to stage II of the motor learning process (Sasaki and Gemba, 1986) (see Fig. 4). The effects of cooling the motor and somatosensory cortex were not only reversible by rewarming but were also fairly unchanged in cooling sessions on different days. Effects of cooling the premotor cortex were also cancelled by rewarming, but they faded out on repeated trials, revealing a kind of hysteretic phenomenon. Probably the substituting compensation by some other brain structures might gradually be accumulated there by repeated coolings. In other words, the function of premotor cortex would shift easily to the other structures even during its temporary impairment by local cooling. Such a possible process might contribute, at least in part, to the phenomenon that the associational cortex is often silent in revealing observable symptoms when it has a localized lesion, and could also be one of the factors that has produced controversies in clinical symptoms and their interpretation when it is involved by any lesion.

Acknowledgements

The authors thank Prof. N. Mizuno for his help and advice on histological checking.

References

1. Sasaki K (1979) Cerebro-cerebellar interconnections in cats and monkeys. In: Massion J, Sasaki K (eds) Cerebro-cerebellar interactions. Elsevier/North-Holland, Amsterdam, pp 105–124
2. Sasaki K, Gemba H (1982) Development and change of cortical field potentials during learning processes of visually initiated hand movements in the monkey. Exp Brain Res 48: 429–437
3. Sasaki K, Gemba H (1983) Learning of fast and stable hand movement and cerebro-cerebellar interactions in the monkey. Brain Res 277: 41–46
4. Sasaki K, Gemba H (1984 a) Compensatory motor function of the somatosensory cortex for the motor cortex temporarily impaired by cooling in the monkey. Exp Brain Res 55: 60–68
5. Sasaki K, Gemba H (1984 b) Compensatory motor function of the somatosensory cortex for dysfunction of the motor cortex following cerebellar hemispherectomy in the monkey. Exp Brain Res 56: 532–538
6. Sasaki K, Gemba H (1986) Effects of premotor cortex cooling upon visually initiated hand movements in the monkey. Brain Res 374: 278–286
7. Sasaki K, Gemba H, Mizuno N (1982) Cortical field potentials preceding visually initiated hand movements and cerebellar actions in the monkey. Exp Brain Res 46: 29–36
8. Sasaki K, Kawaguchi S, Oka H et al (1976) Electrophysiological studies on the cerebello-cerebral projections in monkeys. Exp Brain Res 24: 495–507

Correspondence: Dr. K. Sasaki, Institute for Brain Research, Faculty of Medicine, Kyoto University, Yoshidakonoe-cho, Sakyo-ku, Kyoto 606, Japan.

Acta Neurochirurgica, Suppl. 41, 29–40 (1987)

Plasticity of the Brain in Respect of Functional Restoration After Subarachnoid Haemorrhage

E. Pásztor and J. Vajda

National Institute of Neurosurgery, Budapest, Hungary

Summary

Subarachnoid haemorrhage caused by aneurysmal rupture constitutes a great impact on the brain and on the intracranial content as a whole, with emphasis on the subarachnoid spaces and arteries. The rupture is followed by a wide range of pathological alterations in the neural function and an outcome varying from neglected signs subsiding in a few days to immediate death. Two main factors seem to influence the different events after subarachnoid bleeding. One is the rupture itself which can be extremely variable in severity and in its immediate as well as late consequences. The other is the ability of all parts of the intracranial content to recover. In order to understand either of both the other should also be looked at and both have to be dealt with if we are to treat patients with an aneurysmal rupture properly. For this reason a grading of rupture will be given in respect of some characteristic events in the light of neural restoration. Clearing of CSF, resolution of brain oedema, restoration of impaired CBF, absorption of cisternal and parenchymal haematoma are all of importance. The majority of lesions which developed after the rupture are not fatal or irreversible and even the neural tissue destroyed by the impact or late ischaemia can be functionally replaced. Possible methods of treatment for attaining this functional restoration will be discussed.

Keywords: Subarachnoid haemorrhage; cerebral aneurysma; functional recovery.

A number of question on subarachnoid haemorrhage (SAH) are still open for explanation even today when booming interest and efforts seem to revolutionize the treatment of patients with aneurysmal rupture. Our aim is to analyse the restoration or compensatory processes of symptoms and neurological signs in different periods and stages of SAH, no matter whether the restoration was spontaneous, or the result of any kind of therapy.

With the exception of space-occupying lobar haematomas, surgery for ruptured intracranial aneurysm has long been considered as aiming to prevent rerupture.

Besides, an act of prevention is only justifiable when the clinical state is good enough or an acceptable outcome is expected. Therefore, a lot of the work put in aneurysm-surgery all over the world expresses not only the aim of eliminating the possibility of lethal rebleeding but implies the opinion that many consequences of the first rupture can be reversible, at least in terms of neural function. Going by the title we have to disregard the functional rehabilitation which purely follows surgery for aneurysm since the operation, except these days in some centres, does not target lesions developed as sequelae of the previous rupture. Restoration of function after SAH is a complex, many-sided process and it is not explored in the literature apart from sporadic case reports. According to our experience we think that optimal recovery from SAH and its aftermath depends primarily on the very pathological event which developed during or as a result of the aneurysmal bleeding.

Long-lasting slow-moving recovery after ischaemic brain lesions following SAH can be established by:

1. Progressive recovery when cells are in functional arrest, but not mechanically disrupted.

2. Reorientation of brain circuit to replace lost functions.

3. Neural regeneration (extremely doubtful).

We will review here the different phenomena which occur at aneurysmal rupture and the way the CNS is able to overcome them.

It seems better here not to take notice of subarachnoid haemorrhage from other causes since discussing a number of groups consisting a few individuals would lead to the insignificant.

Which are the most common and universal consequences of bleeding from an aneurysm?

In 8 per cent in our own-, nearly, or over 50 per cent in some Japanese material the haemorrhage (if it was such, of course, was never ascertained) is so much without sequelae that the patient did not see a doctor or when he did, the doctor did not attach great importance to it and the diagnosis was not established: that is the warning sign, or "neglected SAH".

Warning Signs, "Neglected SAH"

Gillingham (1967) drew attention to the "warning leak" by which he meant an initial minor bleed preceding the more severe SAH. Okawara (1973) found warning signs in 48 per cent of his cases, Waga (1975) did so in 59 per cent. The most common symptoms are: headache, nausea, "sprained" neck, neck or back pain. Because of the rather transitory and not threatening character of these symptoms, they are often neglected by the patient or the relatives, moreover, even the physician can miss the significance of these symptoms. Data show (Okawara, 1973; Waga et al., 1975) that warning signs are more common in women, their incidence is higher in cases with an aneurysm on the carotid artery or with multiple aneurysms, while it is lower as age advances. There are also data revealing that the average interval between the warning sign and major attack is around ten days with a shorter delay in cases of carotid and middle cerebral aneurysms (6–7 days).

Okawara widened the term "warning signs" by including all symptoms caused by expansive aneurysms. He found the average interval between these symptoms and the first recognized rupture to be 101 days.

It seems convincing that most of the warning signs are caused by sentinel leaks from the aneurysm (Kassel and Drake, 1982). The quite remarkable observations of Nornes and Magnaes confirm this as they measured ICP continuously and registered peaks of 1,000–2,150 mm H_2O lasting 5–20 min synchronous with the occurrence of clinically evident warning episodes. Surprisingly enough, clear CSF in the basal cisterns at emergency operation in two such cases excluded recent bleeding which could have acted behind these episodes.

Since warning signs are without neurological consequences, they may be called "neglected SAH": this term would express plainly the responsibility of physicians who first see these patients. The striking correlation between the pre-operative state of the SAH patient and the post-operative results seen almost uniformly in all reported material underlines the importance of clear recognition of the warning signs. Some go further by saying that even rebleeding occurs three times more often in poor condition cases (Drake, 1981). It seems evident, that warning signs mean, something which does not cause irreversible brain damage, therefore it is also obvious that patients operated on between the occurrence of warning sign and the supposed first bleeding from the aneurysm have a better chance of a good recovery, almost as good as in cases of silent aneurysm. Furthermore, surgery here does a lot more by grasping the sac in its "active phase".

Grading

There is some correlation between the state of patients with aneurysmal rupture when first seen by the doctor and the severity of rupture, but there are many surprising exceptions which warn us not to draw definite conclusions from the original event of the rupture. The pathological consequences, *i.e.* occasionally a surprisingly large amount of clot in the basal cisterns, rather widespread in certain cases, intracerebral in a few others, are there, and will be responsible for many even lethal consequences if these factors become active in the days after the SAH. We suggest a grading of aneurysmal rupture into four categories of severity:

I. is characterized by a sudden onset of headache with its well-known resemblance to lightening but without loss of consciousness for even minutes.

II. usually starts in the same way but is accompanied by a shorter (less than an hour) period of unconsciousness and full recovery of all functions.

III. This grade consists of patients in whom the rupture has caused longer periods of unconsciousness or marked neurological signs which could be attributed to considerable intraparenchymal bleeding. Patients with shorter unconsciousness or mild focal symptoms but with epilepsy or vegetative symptoms may be graded into this group.

IV. The last grade of rupture includes those who exhibit deep unconsciousness, arrest of some vegetative functions regulated from the brain stem, or unconsciousness for more than a day.

Of course, this grading immediately creates its exceptions, no to mention, that patients who have just suffered SAH with a spell of unconsciousness sometimes cannot recall the loss of consciousness and the doctor puts the case into the mildest group.

The fact that an overwhelming majority of patients suffered SAH grade I–II shows that the original lesions developed at an aneurysmal rupture can resolve spontaneously in those cases who reach hospital and would be suitable for preventive surgery.

Characteristic Forms of Rupture

The obvious clinical pictures of the aneurysmal rupture, however, vary a lot and considering their aftermath they are quite different. The rupture may be preceded by unusual peak of systemic blood pressure or it may follow a sudden drop of severe rise of intra-abdominal or intrathoracic (that is intracranial) pressure, both of which involve an impact on the aneurysm wall by the sudden rise of the transmural pressure. That condition may be combined with areas of various size and weakness on the fundus which after all may face the brain surface or just lie free in the CSF space.

Let us consider some characteristic events of such ruptures in the light of restoration of neural function.

1. The fundus faces the brain substance, partly embedded, the leak is limited, and without abnormal blood pressure: in cases like this there will not much blood spread into the basal cisterns, the brain covers the fissure of the sac, and because of the almost normal blood pressure and the limited size of the slit, no lobar haematoma will develop. These are the ruptures responsible for some of the warning signs, while others, the less modest ones cause the clinical picture of slight SAH, each of which will be easily overcome by the patient without further problems.

What does an aneurysmal rupture of this kind mean to the patients?

A good many of them show no period of unconsciousness. They experience a sudden bursting attack of headache which they elaborate and relate with their individual character. Some also experience latent forms of focal neurological syndromes which subside in minutes or hours. These symptoms by and large have some correlation with this location of the sac but they have no diagnostic force. Their occurrence and fading away are now known to be caused by a spell of benign brain oedema in connection with the vasoparalytic reaction to the sudden intracranial hypertension.

First the elevation of the intracranial pressure by intracranial extravascular blood-volume loading to the intracranial space comes into action (Becker et al., 1975). Both clinical investigation of Nornes and Magnaes (1972), and our animal experiments (Pásztor et al., 1986; Fedina et al., 1986) show that a steep rise of

ICP can be recorded at the time of SAH. The ICP approaches the mean arterial pressure, that means the cerebral perfusion pressure gets depressed toward zero for at least some seconds or minutes. During this period, the CBF exhibits profound alteration into pathological values, probably both in terms of exact flow and its direction. It seems very likely, that for temporary bypass, arteriovenous channels at small arteriolar level begin to function. Assuming the resulted deficit of cerebral blood flow, particularly in regions such as the hypothalamus and the brain stem, and we suppose the hippocampal system, is the primary cause of the period of unconsciousness. All the mentioned investigations accord, however, in demonstrating that the ICP quickly return to normal, the cerebral perfusion pressure increases, and the blood flow deficit is transient enough in all structures not to cause any definite structural damage to the brain even at cellular level, therefore the recovery is rapid and complete.

CSF Clear-up

Other symptoms occurring after SAH, such as headache, neck stiffness, pains in the neck and the back, or sciatica, are the results of haemorrhagic irritation of the meningeal structures. These complaints usually disappear with the clearance of the CSF. The process of clearing CSF, according to Tourtelotte et al. (1964) is not influenced by the number of red cells in CSF. Interestingly enough the ventriculospinal wash out helps to eliminate red cells from CSF only in the first 24 hours after bleeding (McQueen et al., 1974). The CSF clears up within one week or so. One part of red cells gets back to the blood circulation via fila olfactoria and spinal posterior roots through the lymphatic system. Others can be found in the arachnoid villi, which may be in some cases obstructed by the amount of red cells. It is suggested (Ellington and Margolis, 1969; Julow et al., 1979), that the fibrosis and siderosis of the villi may cause chronic intracranial hypertension in some cases of SAH in the later period. A significant quantity of red cells are destroyed by macrophages, that is erythrophagocytosis. The xanthochromia appears very early, about 12 hours after the haemorrhage and peaks on the fifth day or so. The haemoglobin, bilirubin and a negligible amount of methhaemoglobin cause CSF xanthochromia. The conversion of haemoglobin into bilirubin is not a simple process. Experiments by Roost et al. in 1972 showed, that *in vitro* incubated red cells taken from the CSF failed to produce bilirubin, *i.e.* it needed other components, presumably enzyme systems, to participate. They also showed that it is the haemo-oxygenase system responsible for this conversion, and this enzyme could be identified in the arachnoid, choroid plexus and also in cerebral cortex in the rat.

In other cases, the CSF is very reluctant to clear and, it could take more than three weeks. In some of such cases repeated aneurysmal leakage or extensive brain necrosis were found (Tourtelotte et al., 1964). The same slow clearing can be expected in elderly, especially if combined with vascular lesions or permanent neurological deficits.

We may conclude that the mild and moderate forms of aneurysmal rupture evokes various negative events, but they all can subside spontaneously.

Another characteristic combination of factors influencing the event of rupture:

2. The aneurysm lies in the CSF space and rupture is in response to systemic hypertension with a large slit developed on the fundus. This frequent combination of factors is bound to lead a large accumulation of blood in the basal cisterns. The rise of ICP is higher than what we described above, and dominates the first hours. There is a longer period of unconsciousness as a result of lasting intracranial hypertension and more severe reactive brain oedema. Many of these consequences in the clinical picture, however, spontaneously return to normal.

Looking at the clinical state of these patients on admission, it has been proved, that the vast majority of cases with a rupture of grade I–II, are in reasonably good condition during the first two days and the same can be seen in a surprisingly large number of cases of rupture grade III. As a rule, recovery from aneurysmal rupture itself is expected to be as quick as the rupture was mild, but recovery may be complete in any individual. Patients admitted to hospital on the day of rupture often show some disorientation, and an amnesia for the event. The disorientation may last longer and suggests a longer period of acute brain oedema, vasoparalytic reaction to the larger amount of raised intracranial volume and pressure. When the blood filled up the cisterns and the haemorrhage stopped, the intracranial volume-pressure balance began slowly to return to normal. The part of the blood clot peripheral from the aneurysm and in contact with the CSF, is now going to be eliminated as was described above. All consequences of the spell of intracranial hypertension therefore, can subside.

The widespread cisternal blood can cause an acute immediate vasospasm of reactive nature which,

although it lasts only an hour or so, can be significant.

Beyond temporary loss of consciousness and subsequent complaints, meningeal symptoms, there are focal signs, such as hemiparesis and/or aphasia, other hemisphere lesions, which are the most conspicuous neurological changes in such cases. The cause of these symptoms may vary and can be very complex, and sometimes undeterminable or inexplicable. As an example, a total hemiplegia can subside spontaneously in one or two days after SAH, as seen by us in rare cases.

Brain Oedema

Brain oedema can be simply defined as an increase in the water content of the brain tissue which leads to the increase of its volume as a whole. Since randomized clinical studies of various therapeutic regimes have only recently been undertaken, the clinician is overwhelmed with diverging animal experimental data. One desperately tries to find connections between various animal models and the clinical situation of SAH, or to correlate molecular data with the patient's state.

Continuous measurements showed a progressive elevation of ICP following its drop within minutes of the rupture. This period is characteristic in the fourth—fifth hours and is probably caused by the onset of brain oedema. Formation of brain oedema is also a complex process. Dóczi demonstrated that pathological contrast enhancement can be registered on CT in two fifths of patients in the acute stage of SAH (1985), which could be interpreted as the contrast material extravasated as a result of the disruption of the blood-brain barrier. This theory was supported by an animal SAH model showing vasoactive breakdown products causing: 1. an increase of brain water, and sodium content; 2. pathological extravasation of albumin in deeper cortical layers. Disruption of the blood-brain barrier has a characteristic appearance under the electronmicroscope: tight junctions of the capillary walls are open; pinocytosis seems increased in the cytoplasm of endothelial cells, and the mitochondrium here and also in the neuropil is swollen (Klatzo, 1967; Joó et al., 1975; Nagy et al., 1979).

Earlier data of experimental cold lesions, focal ischaemic lesions and intracerebral haematoma revealed that induced arterial hypertension can enormously enhance brain oedema. Recent experimental studies of Hatashita et al. (1986) demonstrated, that focal oedema developed even in normal brain after acute and short-lasting (30 min) non-pharmacologically induced arterial hypertension. The focal oedema was actually located symmetrically in the cortical area of the parieto-occipital arterial boundary zones. In this part of the cortex the widening of extracellular spaces with slight swelling of the astrocytic feet could be observed by histological means. Forty-eight hours after the hypertensive attack, brain oedema did not continue to develop.

After SAH, it is often possible to measure elevated arterial pressure. That is also supported by experimental models (Pásztor et al., 1986). It seems obvious that brain oedema may develop on transient systemic hypertensive basis. Furthermore, systemic hypertensive disease is quite common in patients with aneurysmal rupture.

CT data, however, disagree with the above mentioned, as one can rarely show a localized hypodense area on CT shortly after SAH. If there is a hypodense region on the CT, it correlates with focal hypoxia resulted from vasospasm or thromboembolism.

CT has become an indispensable tool in detecting brain oedema. This method is capable of distinguishing minor variations of linear attenuation coefficients which has led to the possibility of direct imaging of brain oedema (Drayer and Rosenbaum, 1979). Beyond the exact location of the oedema, which is already of paramount importance influencing surgical strategy, it has been also revealed (Cowley, 1983), that the oedema fluid does not spread in a random diffuse way, but follows the course of association bundles. This helps to explain the phenomenon of distant symptoms accompanying focal alterations in SAH. Commissural connections, however, are resistent to the transmission of oedema from remote lesions.

Neurological consequences of brain oedema are benign in nature. Severe paresis, or aphasia can ease or fade away with restoration of all functions if the process of oedema did not last long enough to ruin cell structures. Aggressive anti-oedema therapy is effective and acts rapidly.

The high pressure injected cisternal clot which is already separated from the CSF flow and thus cannot be washed out, begins to act and initiates vasospasm.

Cerebral Vasospasm

One of the most important factors influencing the outcome and prognosis of patients with rupture of intracranial aneurysm is the development and severity of delayed ischaemic neurological lesions caused by vasospasm. It occurs in a disappointingly high percentage of cases and, even more disappointingly, its exact cause or all underlying factors are still unknown.

Since the early days of operative treatment of cerebral aneurysms, neurosurgeons have always known that not a few of the patients exhibit an almost irreversible clinical deterioration some time after operation. Later it became evident, that similar delayed deterioration might occur after the rupture even without any surgical burden on the patient. Severe characteristics spastic stenosis on the main cerebral basal arteries seen on angiograms taken in this period helped to correlate the clinical (ischaemic type) impairment with the process of vasospasm. Countless careful observations, however, disclosed that the same severe appearance of vessels on angiography could be seen in patients examined in a similar time-range after SAH or operation, but they were in excellent or satisfactory condition. Angiography on the other hand, taken in patients with delayed deterioration either after SAH alone or operation for ruptured aneurysm sometimes showed arteries of normal diameter throughout, although this last situation was less common in the clinical practice.

Some argue that cases with vasospasm on angiography and in good condition, have an excellent collateral (pial) circulation in function. Others stated that deteriorated patients without angiographic spasm are likely to have suffered from thrombosis of precapillary vessels not seen on angiograms. It seems convincing, after all, that delayed neurological deterioration of this type attributable to vasospasm, has to be seen as in ischaemic event correlated with CBF disturbances, but probably at a quite different level and special location of vessels. The pressure of the blood burst from the sac, the amount of blood surrounding cerebral arteries may play a tremendous role. Our clinical experience revealed that the morphlogical appearance of narrow vessels on angiograms does not necessarily reflect the clinical symptoms. The morphological detection of these ischaemic lesions of the brain related to vasospasm might be more precisely possible with the advent of new generations of CT. In view of this, we have recently abandoned control angiography for detecting vasospasm apart from CT. The development of delayed ischaemic deterioration, in our practice, is considered on a clinical basis.

More recently, MRI has come up as a promising imaging technique to be used for revealing ischaemic brain areas. The picture provided by MRI is essentially a density map of the mobile protons of tissue water and lipids. Kato et al. (1985) found a significant correlation between the water content of the ischaemic brain and the intrinsic parameters of magnetic resonance spectrometry (T_1 and T_2 relaxation times) under experimental conditions. Abnormalities on MRI corresponded well with various retrospective histochemical pictures: in this way it has become possible to distinguish brain oedema (water accumulation) from severe structural damage injuries of the cellular membrane (ATPase) and in the mitochondria (cytochrome oxidase).

Clinical physiological studies have also helped in understanding the ischaemic process which may follow SAH. Conduction time of somatosensory evoked potential (SSEP) in the affected hemisphere showed significant prolongation pre-operatively in all but patients of Hunt-Hess grade I (Symon, 1985). In the postoperative period a slight but significant prolongation of central conduction time was observed without obligatory deterioration in patients of grade I. Patients of other Hunt-Hess grades showed no change of their conduction time already slightly prolonged preoperatively, unless there was the development of clinical neurological deficit of ischaemic origin.

Recent clinical investigations by Rosenstein et al. (1985) demonstrated that central conduction time is normal with hemispheric blood flow down to the level of 30 ml/100 g/min.

CBF studies demonstrated that as the flow decreases of the level of 15–20 ml/100 g/min, electrical dysfunction of neurons can be detected. Structural damage of these cells usually occurs with a flow below 15 ml/100 g/min (Symon et al., 1975; Sundt et al., 1981; Trojaborg and Boysen, 1973). It seemed also evident from a wider range of studies that there has to be an electrical threshold, i.e. arrest of cerebral electrical activity which was distinguishable from an ionic threshold at which progressive depletion of energy reserves, depolarization of membranes and accumulation of intracellular calcium take place. At a certain time, the process goes over the lethal threshold. The aid of rCBF in these cases seems doubtful, however, as return of neural functions usually precedes rather than follows the increase in rCBF, and early baseline values of rCBF do not predict clinical outcome from acute cerebral ischaemia (Burke et al., 1986).

The sensitivity of different brain structures to ischaemia, and the recovery of impaired functions were also analysed by measuring electrical central conduction time both experimentally and under clinical conditions (Hargadin et al., 1980; Symon, 1985). It was proven, that as one descends in the neuraxis, an increasing resistance of electrophysiological functions to arterial hypotension, that is a decreasing threshold to focal ischaemia, can be demonstrated.

The extent and mechanism of normalization of electrical conduction time during the recovery process after SAH have not yet been sufficiently analysed.

The plasma membrane ion pump system becomes inefficient after ischaemia, which directly leads to the passive exchange of intracellular potassium and extracellular sodium. The resulted decreased neurotransmitter amino acid uptake evokes increased synaptic concentration of glutamate and aspartate with enhanced level of GABA, because re-uptake for both two excitatory neurotransmitters is more sensitive to ischaemia than that for GABA. The process in which these transmitters cause prolonged depolarization of receptor cells all over the brain, and therefore enhance the release of further transmitters, leads to a positive feed-back cascade reinforcement of further glutamate and aspartate release everywhere in a concentration that is neurotoxic. Vizi (personal communication) proved experimentally that pentobarbital and higher Mg^{++} presence inhibit the glutamate release and are therefore quite beneficial in the treatment of hypoxia.

The clinical diagnosis of vasospasm is based on the following factors:

1. Time elapsed from SAH: characteristic between the fourth and ninth days.

2. The rather speedy development of deficits: in hours.

3. The nature of neurological deficits:
a) impaired orientation and consciousness,
b) focal deficits.

Of course, it has to be proved that no other cause of these symptoms (surgical failure, haematoma, thrombosis of vessels) exists (Kassel et al., 1980).

Of all patients 40–70 per cent may develop arterial narrowing seen on angiography after SAH (from the third day onwards reaching a peak at the fifth–ninth days, and with a slow decline of incidence up to the third week). Only 20–30 per cent of all patients develop temporary or permanent delayed neurological deficit (Post et al., 1977).

Experimental evidence shows that after blood flow restoration not only the electrical but ionic-homeostatic mechanisms can return to normal (Hossmann and Kleihues, 1973; Branston et al., 1976; Spetzler et al., 1980). The time necessary for disappearance of symptoms seemed, however, to be influenced not only by the length of the period under which the vessels were occluded but the type of occlusion too, namely much slower recovery was found when similarly long occlusion was applied in a form of repeated on-off occlusions (Spetzler et al., 1980; Weinstein et al., 1986). This cumulative clinical and pathological effect of occlusions of cerebral vessels may be explained by the alteration of functional thresholds on the time line (Spetzler: recruitment response). More recently Hossmann and Ophoff (1986 a, b) showed that in spite of a normal EEG after total cerebral ischaemia lasting one hour there was only partial reappearance of the evoked potentials in the monkey. The protein synthesis gradually approached its normal level in the cortex by the end of the first day, but incomplete recovery was recorded in cortical layer 5 in the border zones. Morphological changes were parallel with lesions in the protein synthesis.

In clinical practice it is surely the ischaemic symptoms of spastic origin, which dominate the picture and determine the functional outcome. Numerous observations show how peculiar the functional recovery of these patients is if they survive the climax of clinical vasospasm. Since the progress of delayed deterioration attributable to vasospasm may be reshaped by an early initiation of aggressive therapy in many cases, we have a scale of the vasospastic symptoms as a marker from which the recovery to a definite clinical state starts.

In case the ischaemia was starting with hemiparesis and the accompanying lobar oedema is modest enough not to dominate the picture, the motor function improvement depends primarily on the extent of the paresis itself, and on the time elapsed from the commencement of the recovery. If the hemiparesis shows a strong cortical character, that is without marked triflexor tone, the improvement is greater and takes much less time. Furthermore it depends greatly on the intact consciousness, mental vitality, and willingness of the patient. Black et al. (1975) demonstrated in monkeys with cortical motor lesion, that active postoperative training of the weak hand alone, or of the weak and normal hands combined, resulted in a higher level of recovery in the weak hand than did retraining of the normal hand alone. Combined training of the weak and normal hand was no better than retraining of the weak hand alone. This suggests that the critical factor in promoting recovery is training of the weak limb.

In cases with basal ganglia involvement, there is a dominance of tonic disorders and their improvement, particularly with lasting lesions, is rather slow and almost never complete. That the recovery in these cases becomes difficult is perhaps due to the impairment of the activity especially in the dominant hemisphere.

After the definite state the redeveloped functions may widen further, but this is much influenced by the quality of the pre-SAH intellect and the possibilities of the roundabout motor solutions, which will primarily be found by the active patient, although rehabilitation surely plays a great role too.

Similarly the age of the patient is a fundamental factor. Chances of functional restitution show a redoubled reduction with increasing age.

It is interesting that although the clinical picture of vasospasm is much bleaker if the vigilance is impaired since it makes all other dysfunctions worse (i.e. otherwise slight dysphasia seems total aphasia), the recovery of these patients is spectacular. Our experience makes us expect more essential restitution of functions in patients with impaired consciousness if unaccompanied by motor symptoms as far as the completeness is concerned, than in cases with acceptable level of wakefulness and severe motor lesions.

Assessment of the definite state is, of course, influenced by the personal qualities of the patient. The term: work in the previous job is a widely used one in the literature, however deceiving, since a patient with a sophisticated job needs quite another restoration of functions compared with the one who works under simple conditions. On the other hand, even a 75 per cent motor restoration is insufficient for a case who earns his living by working with his limbs. Manual workers in an automatic job can afford a slight frontal lesion possessing complete motor abilities, but intellectual workers would rather accept a slight hemiparesis.

Even the long-lasting vegetative state may end up in a satisfactory cure, therefore the confidence put into restoration is of paramount importance in the management of patients with clinical vasospasm.

3. Finally the most dangerous combination of more serious factors: the rupture is evoked by high transmural pressure and a large area of wall defect becomes open where it faces the brain or is embedded in it. This implies that an intracerebral haematoma may develop.

Nornes and Magnaes also took courage in 1972 to measure ICP continuously in patients with aneurysmal rupture, who remained in a severe state. These records showed the instant peak-like increase of ICP followed by a high pressure plateau. In all of these patients an intracranial haematoma could be revealed. There was a significant drop in ICP only when the arterial pressure began to fall as the terminal state commenced in these patients.

The overall mortality of SAH-cases with intracerebral haematoma was much higher, 38 per cent, in a recent multicentre study (Wheelock et al., 1983). Most of these cases were in Hunt-Hess grades IV–V after rupture. The aneurysm responsible for the intracerebral haematoma was on the middle cerebral artery in 45 per cent, and on the anterior communicating artery in 25 per cent. There was a haematoma in the temporal lobe in 46 per cent, and in the frontal lobe in 43 per cent. Overall surgical mortality was as high as 33 per cent in this material, although three quarters of patients underwent only evacuation of the haematoma. Early operation did slightly better in severely ill haematoma-patients since 34 per cent of them have had good recovery. There is some controversy in the interpretation of intracerebral haematoma after aneurysmal rupture and this has importance in the way of outcome. Some authors may regard thick Sylvian fissure clot as a haematoma while others would refer to it as a cisternal accumulation of blood. Most of the intraparenchymal haematomas, if they are not space-occupying, can wait for spontaneous dissolution. Apart from evidence gained from continuous CT observations, we found the haematoma disappeared in many patients with typical CT hyperdensity who had to wait for operation. Haematomas spreading along the association bundles tend to resolve allowing the arrested neural functions to return. Areas and functions show, however, a great difference in their tolerance of non-destructive haematomas. As a rule, the higher is the integration of the function concerned, the wider is the gap between the original and final abilities.

As a rule, an intracerebral haematoma can rupture into the ventricle. Another multicentre study (Mohr et al., 1983) went into the details of patients suffered a SAH which had burst into the ventricle. They found a 64 per cent mortality among these severely ill cases. The primary location of aneurysm causing an intraventricular haematoma was, quite predictably the anterior communicating artery (40 per cent), and that was followed by the carotid (25 per cent).

The width of ventricles filled with blood had a much greater influence on the outcome of these patients than the actual size of the clot itself, as no patient with a ventriculo-cranial ratio of over 0.25 on the initial CT survived.

In the era of regular use of CT in cases of SAH the incidence of recognized intraventricular clot increased significantly (retrograde bleeding through the foraminal cisterns into caudal ventricles has to be excluded). Patients who easily survive an accumulation of blood within the ventricles are not unique any more. The fundamental difference between the dangerous and

"benign" ventricular clots is that the ventricular tamponade on one hand gives the CSF no way to flow and the intracranial hypertension is made worse by the increase in CSF volume. The moderate type ventricular clot on the other hand is the subject of a continuous process of elimination but it takes longer than those in the basal cisterns, and as a negative effect, it may cause the choroid plexus to increase CSF formation. With a resonably free CSF flow, however, this latter condition settled spontaneously in almost all cases we saw.

There are other serious conditions which constitute a severe threat to life and function after SAH.

Haemorrhagic Infarction

Anemic infarction can quite commonly be seen in the population concerned and this CT and pathological finding can obviously be attributed to the vasospastic process. A small portion (6 per cent) of these infarctions turn to haemorrhagic ones and are characteristic in the fourth week after SAH. Generally it does not increase the mass effect of the inital infarct except in a few cases in whom massive haemorrhage occurs into the infarct (Terada et al., 1986).

This fact has a strong therapeutic implication, as hypertensive therapy is no longer necessary while the remission stage is on after vasospasm, but it can do a lot harm to the patient by making the anaemic area turn to a haemorrhagic one.

Vegetative Disturbances

Cardiac Abnormalities

There is a well-known higher incidence of other systemic vascular diseases such as hypertension, coronary lesion, among SAH patients. Severe acute cardiac abnormalities, however, can accompany aneurysmal rupture. The connection of these events seems to gain further argument from those observations which showed that the suddenly developing new cardiac disturbances disappear simultaneously with the clinical recovery from SAH. In his review on cardiac signs of SAH, Marion (1986) summarized the ECG abnormalities (Q-T prolongation, T wave inversion, S-T segment depression or elevation, various dysrhythmias) and myocardial lesions (subendocardial myocytosis, myofibrillary degeneration). These findings correlate with data of experimental SAH in which there is an increase of catecholamine in the serum after SAH. Our own experiments demonstrated a sympathetic overactivity recordable from three sympathetic nerves (nn. vertebralis, cardiacus, and renalis) simultenaously (Pásztor et al., 1986; Fedina et al., 1986). These (Figs. 1–3) experiments accord with the clinical experiences: propanolol, an autonomic blocker seems to be the most appropriate agent against cardiac abnormalities during the acute phase of SAH.

Respiratory Disturbances

In a few among those who survive the rupture of their intracranial aneurysm, acute brain stem dysfunctions manifest in respiratory irregularities or even failure. Mild or severe disturbances of ventilation have a further impact on the brain by lowering $PaCO_2$ and worsening the energy supply to vital brain areas. Most of these cases have their aneurysm on the lower posterior circulation.

Perspectives of Types of Treatment in Functional Restoration After SAH

Ischaemic Consequences

Millikan (1955) was a pioneer in using a vasodilator agent for treating ischaemic brain lesions. Since these attempts several drugs working with various mechanisms were labelled promising, but general enthusiasm has not yet been developed on any of the medicine used and still in use.

The enormous number of references on drugs and other manipulations (Wilkins 1986, 318 references) such as the original use of carbon dioxide has made us cite only the most recent relevant reviews.

The calcium channel blocker: nimodipine is a drug now under world-wide testing and seems to be able to reverse the ischaemia induced by vasospasm (Meyer et al., 1986). After clinical trials on propanolol which has proved effective in acute myocardial lesions but ineffective enough in preventing cerebral ischaemia, the racemic (d,1)-propanolol has been tested with preliminary beneficial effect (Standefer and Little, 1986). Suzuki and Yoshimoto (1979) earlier, and Jafar's recent clinical investigations (1986) showed the mannitol given 4.5 g/kg weight/day increased the rCBF by half of the initial level on average, and at the same time did not decrease ICP. That means a further advantage, because the bursting pressure against an aneurysm-wall (systolic arterial pressure—ICP) could be kept unchanged which prevented the aneurysms from rerupturing in all 16 patients who were treated with mannitol.

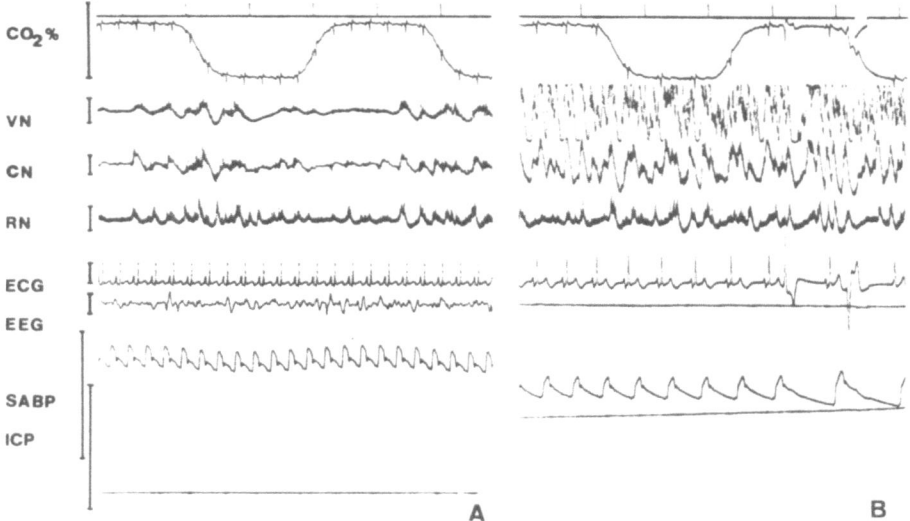

Fig. 1. The effects of the intracisternal injection of 4.0 ml fresh arterial blood are demonstrated on the activity of the peripheral sympathethic nerves, and on the ECG and systemic arterial blood pressure. On Panel A the pre-injection records show the well-known respiratory and pulse-synchronous grouping of the activity of the three sympathethic nerves (n. vertebralis, n. cardiacus, n. renalis). The ECG (lead II) and heart rate is normal (around 210/min), the blood pressure is 170/140 mm Hg. The ICP is 7–8 mm Hg. On Panel B about 60–66 sec after the beginning, *i.e.* during the injection of 4.0 ml blood into the cisterna magna, together with the elevation of the ICP up to 280–320 mm Hg, the sympathetic activity markedly increased in all three nerves. At the same time severe bradycardia with 126/min heart rate and ECG changes with extra systoles of different origin (*e.g.* ventricular) could be observed, and the arterial blood pressure rose to 270/180 mm Hg. In the second half of the Panel B two extra systoles were followed by a peripheral pulse deficit. Abbreviations and calibrations: time in seconds; endtidal CO_2 %: 0–6%; VN, CN, RN (n. vertebralis, n. cardiacus, n. renalis sympathethic nerves): 25-100-25 mV; ECG: 500 mV; EEG: 100 mV, SABP (Systemic Arterial Blood Pressure) in Panel A: 0–200 mm Hg, in Panel B: 0–400 mm Hg; ICP (Intracranial Pressure) in panel A: 0–40 mm Hg, in Panel B: 0–400 mm Hg

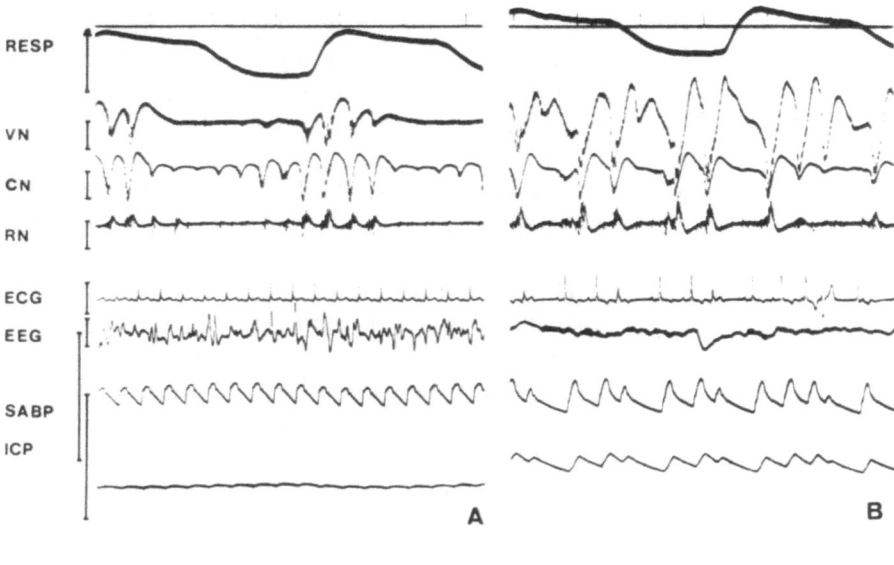

Fig. 2. Similar phenomena are demonstrated as in Fig. 1, but during the intracisternal injection of 3.0 ml mock cerebrospinal fluid. Here again severe brady-arrhythmia could be observed with extra systoles, and the peripheral pulse pressure showed remarkable changes in amplitude. The similarity of the vegetative phenomena during the intracisternal injection of arterial blood, or mock cerebrospinal fluid calls in question the specificity of blood causing these vegetative disturbances in patients suffering from subarachnoid haemorrhage. Abbreviations and calibrations as in Fig. 1

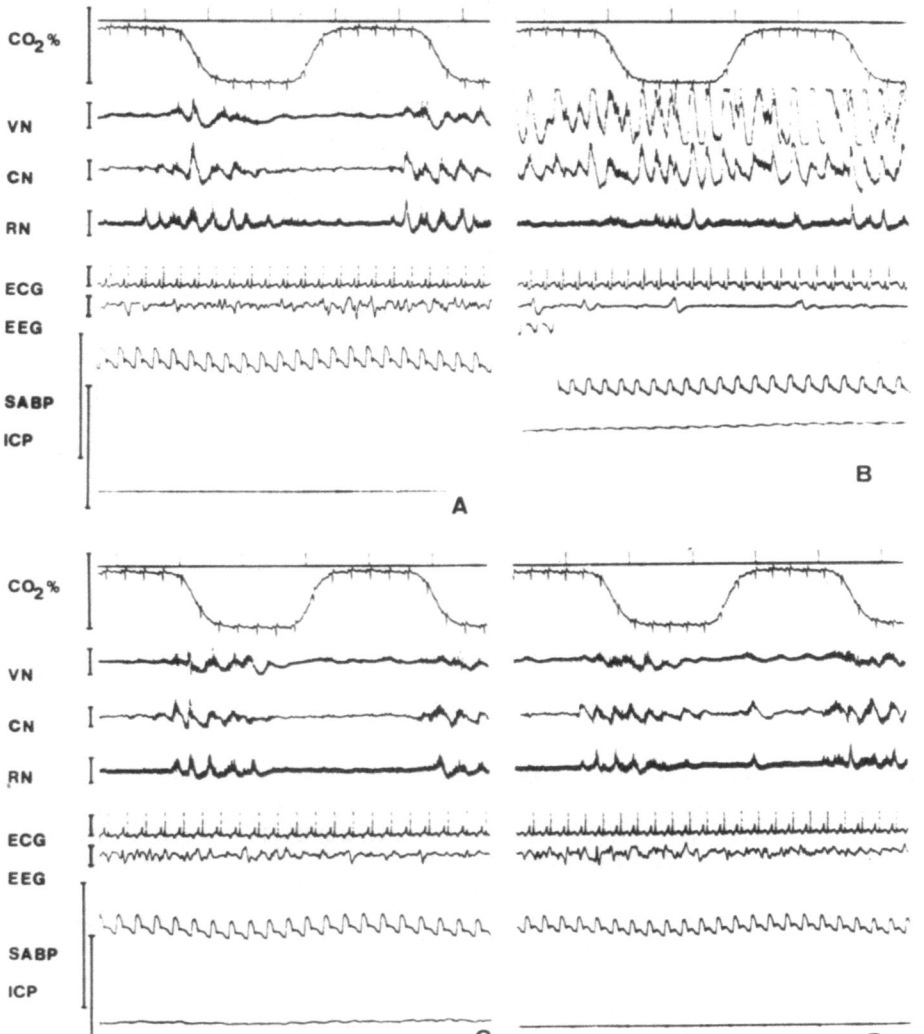

Fig. 3. The effects of intracisternal injection of 4.0 ml fresh arterial blood are demonstrated during the intracranial hypertension, shortly after the end of injection, and at the time of the so-called "early vasospasm" period. The records are taken from the same experiment, as the records of the Fig. 1. Panel A shows the pre-injection period. On Panel B about 48–54 sec after the beginning of the intracisternal injection, the activity of the vertebral and cardiac sympathetic nerves markedly increased, while the activity of the renal sympathethic nerve slightly decreased. At that time mild and short-lasting tachycardia (about 240/min heart rate) and some ECG changes developed (elevation of ST segment), and the blood pressure rose from 170/140 mm Hg up to 250/200 mm Hg. On Panels C and D the records are taken about 2.5 and 10 min after the end of the intracisternal injection. It can be seen on both panels, that the overactivity of the peripheral sympathetic nerves has already diminished while the level and the pattern of the activity is similar to that of the preinjection period. The ECG and heart rate is normal, and the blood pressure is somewhat lower, than in the control state (140/110 mm Hg). Abbreviations as in Fig. 1 but ICP in Panel A: 0–40 mm Hg, in panel B: 0–400 mm Hg, in Panels C–D: 0–100 mm Hg

Michenfelder (1984) reported good results with administering a thiopental bolus and, if haemodynamically tolerated, increased isoflurane concentrations.

Perfluorochemicals and glycerol are now in focus by protecting the neural tissue form ischaemia. ^{32}P-MR-spectrum was used to prove the effectiveness of these drugs (Naruse et al., 1984). The method is also promising for future research. Another newcomer, nizofenon (an imidazole derivation) was named by a multicentre group (Ohta et al., 1986) as beneficial for patients with delayed ischaemia.

The most important development so far in preventing delayed ischaemia attributable to vasospasm has been the widespread acceptance of early operation for ruptured aneurysm. There are now reasons to believe that a comparably large amount of blood burst from the aneurysm with sudden high pressure can distened the basal cisterns, and since the breakdown products of that mass of cisternal blood isolated from still functioning CSF flow cause cerebral arteries to develop vasospasm, an early operation, before fibrotic adhesions from the clot could form, can eliminate the primary source of vasospasm. In our experience, as stated by many, first of all Japanese, authors (Mizukami et al., 1978; Sano and Saito, 1978; Yoshimoto et al., 1979; Suzuki et al., 1979), the possibility of removing the clot from basal cisterns by the gentle method which the vessels require is the greatest on the first day, and acceptable in the majority of the cases within 48 hours. The results of operation within that 48 hours have proven the idea behind it, and also by cutting the incidence of later rebleeding early operation allows more aggressive treatment by induced hypertension, hypervolaemia, another convincing adjunct to the prevention of delayed ischaemia (van der Werf, 1986). Moreover, Muizelaar and Becker emphasized (1986) the importance of slight haemodilution by maintaining the haematocrit around 70 per cent below normal. Grotta et al. (1985) proposed the perfusion of potentially viable ischaemic brain tissue by increasing the CBF and cardiac output, lowering haematocrit and expanding left ventricular and diastolic volume. According to the clinical experiences of Finn et al. (1986) prompt correction of low pulmonary wedge pressure (at 10–12 mm Hg) has proved effective in preventing or reversing neurological deficits following aneurysmal SAH.

In *conclusion*, people who deal with or care for patients who survived the rupture must be aware that the majority of pathological lesions developed at or in consequence of the rupture are not fatal, or irreversible, particularly if treated properly and in time. Even the neural substance destroyed by the impact or late ischaemia, can be functionally replaced. Therefore all efforts have to be made to save SAH patients to an utter end when a proper rehabilitation be worked out. Much research has to be done in the future to reveal methods which are able to distinguish between definite destruction of the neural tissue and the temporary, however long, arrest of function.

References

1. Becker DP, Young HF, Vries JK et al (1975) Monitoring in patients with brain tumours. Clin Neurosurg 22: 364–388
2. Black P, Markowitz RS, Cianci SN (1975) Recovery of motor function after lesions in motor cortex of monkey, in: Outcome of severe damage to the central nervous system, CIBA Foundation Symposium 34. Elsevier, Amsterdam, pp 65–83
3. Branston NM, Symon L, Crockard HA (1976) Recovery of the cortical evoked response following temporary middle cerebral artery occlusion in baboons: Relation to local blood flow and PO_2. Stroke 7: 151–157
4. Burke AM, Younkin D, Gordon J et al (1986) Changes in cerebral blood flow and recovery from acute stroke. Stroke 17: 173–178
5. Cowley AR (1983) Influence of fiber tracts on the CT appearance of cerebral edema: anatomic-pathologic correlation. Am J Neurorad 11: 915–925
6. Dóczi T (1985) The pathogenetic and prognostic significance of blood-brain barrier damage in the acute stage of aneurysmal subarachnoid haemorrhage. Clincal and experimental studies. Acta Neurochir (Wien) 77: 110–132
7. Drake CG (1981) Management of cerebral aneurysm. Stroke 12: 273–283
8. Drayer BD, Rosenbaum A (1979) Brain edema defined by cranial computed tomography. J Comput Assist Tomogr 3: 315–323
9. Ellington E, Margolis G (1969) Block of arachnoid villus by subarachnoid haemorrhage. J Neurosurg 30: 651–657
10. Fedina L, Pásztor E, Kocsis B et al (1986) Activity of peripheral sympathetic efferent nerves in experimental subarachnoid haemorrhage. Part II: Observations during the "early vasospasm" period. Acta Neurochir (Wien) 80: 42–46
11. Finn SS, Stephensen SA, Miller CA et al (1986) Observations on the perioperative management of aneurysmal subarachnoid haemorrhage. J Neurosurg 65: 48–62
12. Gillingham FJ (1967) The management of ruptured intracranial aneurysms. Scott Med J 12: 377–383
13. Grotta JC, Pettigrew LC, Allen S et al (1985) Baseline haemodynamic state and response to hemodilution in patients with acute cerebral ischemia. Stroke 16: 790–795
14. Hargadine JR, Branston NM, Symon L (1980) Central conduction time in primate brain ischemia—a study in baboons. Stroke 11: 637–643
15. Hatashita S, Hoff JT, Ishii S (1986) Focal brain edema associated with acute arterial hypertension. J Neurosurg 64: 643–649
16. Hossmann KA, Kleihues P (1973) Reversibility of ischemic brain damage. Arch Neurol 29: 375–384
17. Hossmann KA, Ophoff BG (1986) Recovery of monkey brain after prolonged ischemia. I. Electrophysiology and brain electrolytes. J Cereb Blood Flow Metab 6: 15–21

18. Hossmann KA, Ophoff BG (1986) Recovery of monkey brain after prolonged ischemia. II. Protein synthesis and morphological alterations. J Cereb Blood Flow Metab 6: 22–23

19. Jafar JJ, Johns LM, Mullan SF (1986) The effect of mannitol on cerebral blood flow. J Neurosurg 64: 754–759

20. Joó F, Rakonczay Z, Wollemann M (1975) cAMP-mediated regulation of the permeability in the brain capillaries. Experimentia 31: 582–583

21. Julow J (1979) Prevention of subarachnoid fibrosis after subarachnoid haemorrhage, with urokinase. Acta Neurochir (Wien) 51: 53–61

22. Kassell NF, Drake CG (1982) Timing of aneurysm surgery. Neurosurgery 10: 514–519

23. Kassell NF, Peerless SJ, Drake CG et al (1980) Treatment of ischemic deficits from cerebral vasospasm with high dose barbiturate therapy. Neurosurgery 7: 593–597

24. Kato H, Kogure K, Ohtomo H et al (1985) Correlations between proton nuclear magnetic resonance imaging and retrospective histochemical images in experimental cerebral infraction. J Cereb Blood Flow Metab 5: 267–274

25. Klatzo J (1967) Neuropathological aspects of brain edema. J Neuropathol Exp Neurol 26: 1–14

26. Marion DW, Segal R, Thompson ME (1986) Subarachnoid haemorrhage and the heart. Neurosurgery 18: 101–106

27. McQueen JD, Northrup BE, Leibrock LG (1974) Arachnoid clearance of red blood cells. J Neurol Neurosurg Psychiat 37: 1316–1321

28. Meyer FB, Anderson RE, Yaksh TL et al (1986) Effect of nimodipine on intracellular brain pH, cortical blood flow, and EEG in experimental focal cerebral ischemia. J Neurosurg 64: 617–626

29. Michenfelder JD (1984) Cerebral preservation for intraoperative focal ischemia. Clin Neurosurg 32: 105–113

30. Millican CH (1955) Evaluation of carbon dioxide inhalation for acute focal cerebral infraction. Arch Neurol 73: 324–328

31. Mizukami M, Takemae T, Kin H et al (1978) Computed tomography of ruptured intracranial aneurysm in acute stage—relationship between vasospasm and high density on CT scan. Brain and Nerve 30: 861–866

32. Mohr G, Ferguson G, Kahn M et al (1983) Intraventricular haemorrhage from ruptured aneurysm. J Neurosurg 58: 482–487

33. Muizelaar JP, Becker DP (1986) Induced hypertension for the treatment of cerebral ischemia after subarachnoid haemorrhage. Direct effect on cerebral blood flow. Surg Neurol 25: 317–325

34. Nagy Z, Mathieson G, Hüttner J (1979) Blood-brain barrier opening to horseradish peroxidase in acute arterial hypertension. Acta Neuropathol 48: 45–53

35. Naruse S, Horikawa Y, Tanaka C et al (1984) Measurements of in vivo energy metabolism in experimental cerebral ischemia using ^{31}P-NMR for the evaluation of protective effects of perfluorochemicals and glycerol. Neurol Res 6: 169–175

36. Nornes H, Magnaes B (1972) Intracranial pressure in patients with ruptured saccular aneurysm. J Neurosurg 36: 537–547

37. Ohta T, Kikuchi H, Hashi K et al (1986) Nizofenone administration in the acute stage following subarachnoid haemorrhage. J Neurosurg 64: 420–426

38. Okawara SH (1973) Warning signs prior to rupture of an intracranial aneurysm. J Neurosurg 38: 575–580

39. Pásztor E, Fedina L, Kocsis B et al (1986) Activity of peripheral sympathetic efferent nerves in experimental subarachnoid haemorrhage. Part I: Observations at the time of intracranial hypertension. Acta Neurochir (Wien) 79: 125–131

40. Post KD, Flamm ES, Goodgold A et al (1977) Ruptured intracranial aneurysms. Case morbidity and mortality. J Neurosurg 46: 290–295

41. Rosenstein J, Wang AD, Symon L et al (1985) Relationship between hemispheral CBF, CCT and clinical grade in aneurysmal subarachnoid haemorrhage. J Neurosurg 62: 25–30

42. Sano K, Saito I (1978) Timing and indication of surgery for ruptured intracranial aneurysms with regard to cerebral vasospasm. Acta Neurochir (Wien) 41: 49–60

43. Spetzler RF, Selman WR, Weinstein P et al (1980) Chronic reversible ischemia: evaluation of a new baboon model. Neurosurgery 7: 257–261

44. Standefer M, Little JR (1986) Improved neurological outcome in experimental focal cerebral ischemia treated with propanolol. Neurosurgery 18: 136–140

45. Sundt TM Jr, Sharbrough FW, Piepgras DG (1981) Correlation of cerebral blood flow and electroencephalographic changes during carotid endarterectomy. Mayo Clin Proc 56: 533–543

46. Suzuki J, Onuma T, Yoshimoto T (1979) Results of early operations on cerebral aneurysms. Surg Neurol 11: 407–412

47. Suzuki J, Yoshimoto T (1979) The effect of mannitol in prolongation of permissible occlusion time of cerebral arteries: Clinical data of aneurysm surgery. In: Suzuki J (ed) Cerebral aneurysms. Neuron Publishing Co, Tokyo, pp 330–337

48. Symon L (1985) Thresholds of ischaemic applied to aneurysm surgery. Acta Neurochir (Wien) 77: 1–7

49. Symon L, Crockard HA, Dorsch NWC et al (1975) Local cerebral blood flow and vascular reactivity in a chronic stable in baboons. Stroke 6: 482–492

50. Terada T, Komai N, Hayashi S et al (1986) Haemorrhagic infarction after vasospasm due to ruptured cerebral aneurysm. Neurosurgery 18: 415–418

51. Tourtellotte WW, Metz LN, de Yong RN (1964) Spontaneous subarachnoid haemorrhage. Neurology 14: 301–309

52. Trojaborg W, Boysen G (1973) Relation between EEG, regional cerebral blood flow and internal carotid artery pressure during carotid endarterectomy. Electroencephal Clin Neurophys 34: 61–69

53. Van der Werf AJM (1986) Spasme vasculaire et ischémie cérébrale après hémorragie par rupture anévrysmale. Neurochirurgie 32: 1–22

54. Waga S, Ohtsuto K, Handa H (1975) Warning signs in intracranial aneurysms. Surg Neurol 3: 15–20

55. Weinstein PR, Anderson GG, Telles DA (1986) Neurological deficit and cerebral infarction after temporary middle cerebral artery occlusion in unanesthetized cats. Stroke 17: 318–324

56. Wheelock B, Weir B, Watts R et al (1983) Timing of surgery for intracerebral hematomas due to aneurysm rupture. J Neurosurg 58: 476–481

57. Wilkins RH (1986) Attempts at prevention or treatment of intracranial arterial spasm: an update. Neurosurgery 18: 808–822

58. Yoshimoto T, Uchida K, Kaneko U et al (1979) An analysis of follow-up results of 1000 intracranial saccular aneurysms with definitive surgical treatment. J Neurosurg 50: 152–157

Correspondence: Dr. E. Pásztor, National Institute of Neurosurgery, Budapest, Hungary.

Acta Neurochirurgica, Suppl. 41, 41–45 (1987)

Transplantation of the Suprachiasmatic Nucleus in the Rat

Y. Saitoh, I. Nihonmatsu, and **H. Kawamura**

Department of Neuroscience, Mitsubishi-Kasei Institute of Life Sciences, Machida-shi, Tokyo, Japan

Summary

Restoration of the circadian rhythmicity in wheel-running activity was shown in rats with bilateral suprachiasmatic nuclei (SCN) lesions, after transplantation of the neonatal SCN into the wall of the third ventricle. Free-running circadian rhythms of the wheel-running activity were recorded in young adult rats at least for a month under constant dark condition. Then, bilateral SCNs were completely lesioned electrolytically under deep pentobarbital anaesthesia.

After further recording for more than two months without obvious circadian rhythmicity in wheel-running activity, the animals were subjected to transplantation of the SCN. SCNs taken from day 1 neonatal rats were transplanted by injecting the grafts into the third ventricle of the host rat under pentobarbital anaesthesia. After recovery from the procedure, the rat was returned to a cage with a running wheel. Food and water were available at all times. Successful transplantation led to restoration of the circadian rhythmicity starting from two weeks and up to three months after the transplantation.

To identify the SCN in the transplanted graft, we used an immunohistochemical staining method for the VIP (vasoactive intestinal polypeptide) and vasopressin. The VIP was located particularly in the ventral area of the SCN, whereas vasopressin was in the dorsal area. In most cases, where circadian rhythmicity was successfully restored, the graft was attached to the caudal wall of the third ventricle.

Keywords: Suprachiasmatic nucleus (SCN); circadian rhythm; vasoactive intestinal polypeptide (VIP); vasopressin.

Introduction

Evidences indicating that the SCN is a potent self-sustained endogenous oscillator generating circadian rhythms have been accumulated in our laboratory (Ibuka and Kawamura, 1975; Ibuka, Inouye and Kawamura, 1977; Kawamura and Inouye, 1979; Inouye and Kawamura, 1979; Inouye and Kawamura, 1982). After surgical isolation of a hypothalamic island containing the SCN *in situ*, neural activity inside the island showed a clear circadian rhythm whereas circadian rhythms elsewhere outside the island which were recorded in intact animals, were totally disrupted.

These findings seem to suggest two important aspects in the physiology of mammalian circadian rhythms. Firstly, a self-sustained oscillator is located in the SCN, and so far no other comparably potent oscillator has been found elsewhere in the rat brain. Secondly, since after such an isolation of the SCN *in situ*, circadian rhythms in various functional and behavioural activities were disrupted, it seems quite likely that circadian rhythms generated in the SCN are not transmitted by humoral means but propagated through neural pathways. The objection that there may be some oscillator outside the brain which affects the SCN activity in the hypothalamic island via blood flow or via cerebrospinal fluid does not seem feasible. The reason why, is that such an evident self-sustained oscillator activity outside the brain has not been found up to now, and endocrine organs are quite unlikely candidates as a generator of the circadian rhythm in mammals (Richter, 1967). Furthermore, a characteristic of the SCN as a quite potent oscillator of the circadian rhythm has been shown by lesion experiments. If one made a complete lesion of the SCN in one side and a quarter or one fifth of the SCN remained on the other side, it was sufficient to induce circadian rhythmicity in behaviour (van den Pol, 1979).

Taking all these findings into consideration, we thought that transplantation of the SCN into arrhythmic rats which had previously received bilateral SCN lesions, might serve as a good example to observe a functional restoration in the brain after a transplantation of its lost part. Preliminary results on successful restoration of circadian ryhthmicity in wheel-running activity after transplantation of hypothalamic tissue containing the SCN have been reported elsewhere (Sawaki and Kawamura, 1983; Kawamura, Nihonmatsu, Saitoh et al., 1984; Sawaki, Nihonmatsu and Kawamura, 1984).

Materials and Methods

Wistar strain male albino rats bred in our Institute were used in these experiments. Wheel-running activity was examined in young male rats (about 2 to 3 months after birth) and control records were taken for more than one month under constant dark conditions. Rats which showed clear free-running circadian rhythms were used for further experiments. Under deep pentobarbital anaesthesia, each rat was placed on a stereotaxic instrument. Electrodes were stereotaxically inserted through a trephine hole in the cranium into the SCN bilaterally and electrolytic lesions of the SCN were made by passing DC current. After recovery from operation, the rat was returned to a home cage with running wheel in a sound-attenuated, light controlled chamber. Food and water were available at all times.

Transplantation was performed in rats which showed arrhythmicity for more than two months in wheel-running activity after SCN lesions. For the transplantation procedure, each rat under deep pentobarbital anaesthesia was mounted on a stereotaxic instrument. An inner tube with an internal diameter of 0.9 mm for tissue transplantation was inserted into the outer guide tube whose outside diameter was about 1.8 mm. The length of the inner tube was adjusted so that the tip of the inner tube came out slightly from the tip of the outer guide tube.

The inner tube was connected with a tuberculin syringe filled with Dulbecco solution. The outer tube was held by an electrode holder of the stereotaxic instrument. The tubes were inserted into the caudal part of the third ventricle. After these procedures, a neonatal (day 1) rat was sacrificed under ether anaesthesia and the brain was quickly removed. Slices of the brain with frontal sections of about 1 mm thickness were made. It was then immediately immersed into chilled Dulbecco solution in a dish. A slice containing the SCN was carefully placed on a small silastic plate immersed in the dish. Using a inner tube of the transplantation instrument, each SCN was punched out and sucked into the inner tube slightly pulling the syringe connected with the inner tube.

The inner tube was then immediately inserted into the guide tube, the tip of which had been already advanced into the third ventricle of the host rat. Usually the inner tube contained a pair of SCNs taken from a rat, but in some cases, four pieces of SCNs taken from two neonatal rats were injected into the third ventricle. After the transplantation, both the guide tube and injection tube were removed. The trephine hole was closed by dental cement and the surgical wound was sutured. Antibiotics and Ringer solution were injected as usual.

Wheel running activity was recorded for more than two months after transplantation. For recording wheel running activity, each full turn of the wheel by the rat in the cage activated a switch which moved a pen of a slow speed pen-recorder. The signal was simultaneously stored in a memory disc which was used for computer processing of the data to write out various diagrams and to make power spectrum analysis.

After completion of the experiment, each rat was injected colchicine (140 µg) into the lateral ventricle. Six to forty-eight hours after injection, the rat was killed with an overdose of pentobarbital. The brain was perfused with saline followed by 10% formalin solution. VIP (vasoactive intestinal polypeptide) and vasopressin were stained using a peroxidase-antiperoxidase (PAP) method with 20 to 30 µm sections.

Results

Fig. 1 shows some sample data taken from a rat (TP 93). A control record taken for four weeks before the SCN lesion is seen at the top on the left. Each line denoted successive 2 days record and the record in the right hand half is repeated in the left hand half of the successive line and so on (double-plotted method). This control record shows a circadian rhythm with a period slightly longer than 24 hours.

After electrolytic lesions of the bilateral SCN (the record is shown below on the left in Fig. 1), the circadian rhythm was totally disrupted. In these records, the number of each full turn of the wheel in every 15 min was accumulated and indicated as the length of a column. Thus, before lesions, 100 turns in 15 min should fill the column up to the top. This was indicated as Max. 100 in the diagram. After lesions, therefore, 50 turns in 15 min would fill a column, and in the right side record, which shows wheel running behaviour in the rat after SCN transplantation, 25 turns would make a full column. This means a four times amplification of the after-transplantation record compared to the pre-lesion record with max. 100 column. In this rat, for about 100 days after transplantation, no obvious circadian rhythm was observed under constant dark condition. However, after 100 days, though weaker compared to the pre-lesion record, a significant circadian rhythm was restored. Apparently, the period of the restored circadian rhythm was shorter than 24 h.

Power spectrum diagrams shown in Fig. 2 were processed from the data stored in the memory disc. Each diagram shown in Fig. 2 was processed using the data taken from the period indicated as a, b–f in Fig. 1. The diagram a denotes control data indicating a significant peak of about 24 h period (circadian rhythm). Diagrams b and c showed no 24 h peak after the SCN lesion. Immediately after transplantation (d), no 24 h peak could be seen but 100 days after, a 24 h peak (circadian rhythm) was restored (e, f). In diagrams b to f, note that the intensity (ordinate) was amplified twice. Fig. 3 shows a large lesion in the basal hypothalamic area which totally eliminated the SCNs. Transplanted grafts were demonstrated by arrows and apparently a large graft shown at the bottom (left) was the one which contained live SCN, because both VIP

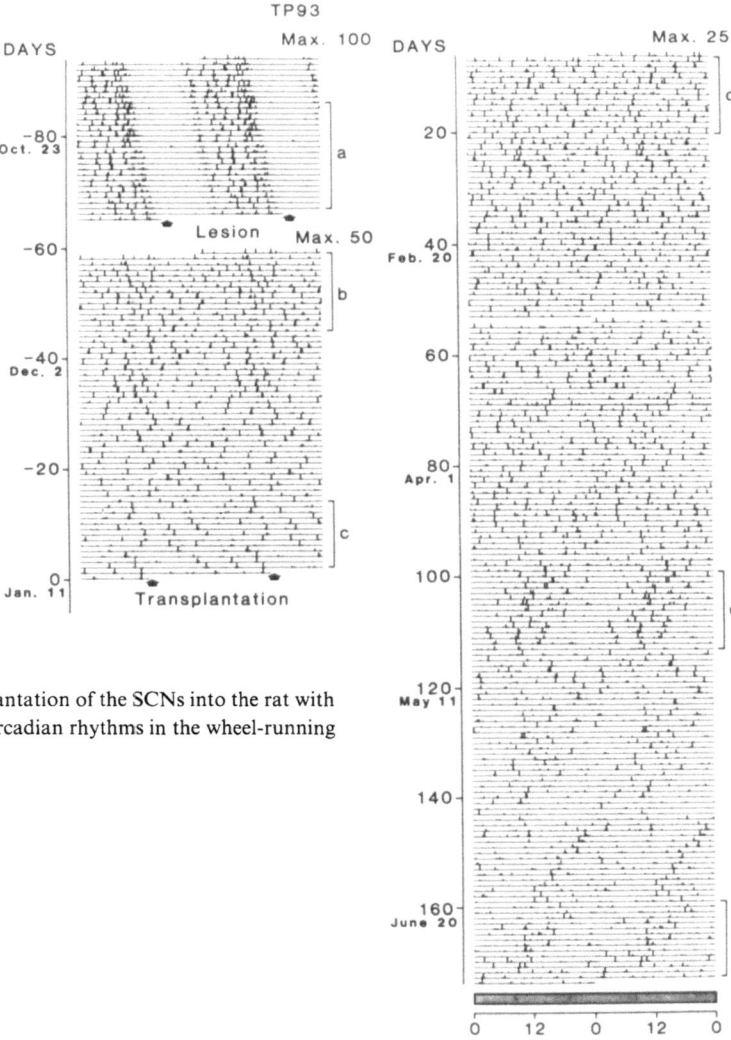

Fig. 1. Restoration of the circadian rhythmicity after transplantation of the SCNs into the rat with bilateral SCN lesions. Note totally different periods of the circadian rhythms in the wheel-running activity before lesion and after transplantation

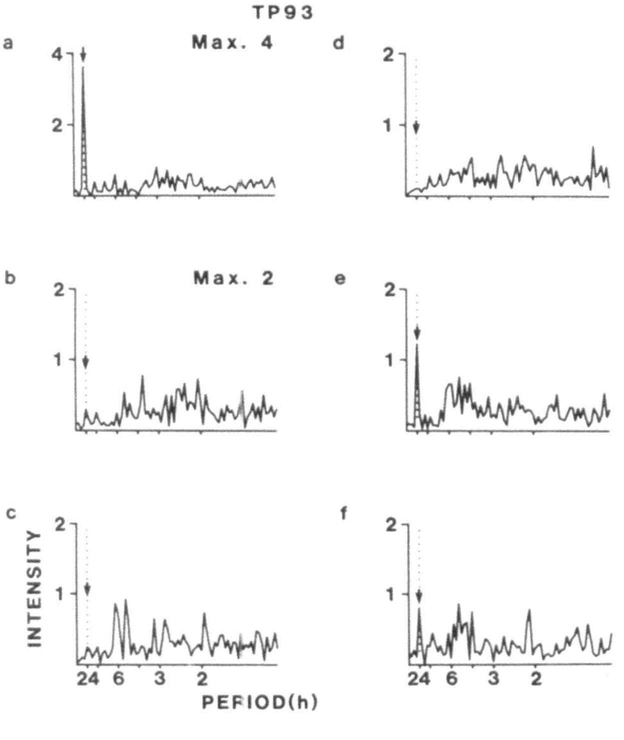

Fig. 2. Power spectrum diagram showing disappearance of about 24 h peak after SCN lesions (b, c) compared to the control intact animal (a). Gradual restoration of circadian rhythmicity after transplantation of the SCN (d, e, f)

TP93

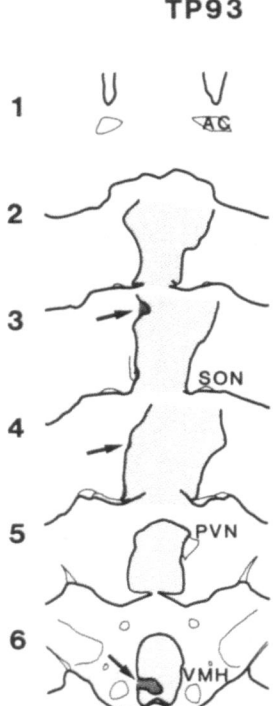

Fig. 3. Diagram indicating a large hypothalamic lesion totally eliminating SCN. Each graft shown with an arrow was covered by the ependyma. Both VIP and vasopressin were found in the graft shown left at the bottom

thalamus contained both VIP and vasopressin with such a density, this method was used to identify SCN in the graft (also see Kawamura and Nihonmatsu, 1985). Dense VIP containing neurons were shown in the bottom graft depicted in Fig. 3, and vasopressin containing neurons were also found in the same graft. Here, only a case of successful transplantation was described. All other available data will be summarized and published elsewhere.

Discussion

Transplantation of the SCN graft to the wall of the host arrhythmic rat with total SCN lesions restored circadian rhythms in wheel-running activity. A question arises, why the SCN transplanted to a different site far from the original SCN could restore the circadian rhythmicity? Such a result could be interpreted as follows. First of all, the SCN itself is quite a potent, self-sustained oscillator of the circadian rhythm. So potent that only a part of it is enough to induce circadian rhythmicity in the whole body. Secondly, it is known that the periventricular area has many neural endings coming from the SCN (Card et al., 1981). This might have some beneficial effect for recovery of the functional connection between the host brain and donor's SCN. The connection of the SCN with the periventricular area may easily affect via this area the reticular activating system, inducing behavioural circadian rhythms. The rostral part of the brain stem activating

and vasopressin containing neurons were clearly found in this graft. For explanation of this finding, Fig. 4 indicates VIP containing neurons located in the ventrolateral area of the SCN. On the other hand, vasopressin containing neurons were located in the dorsomedial area of the SCN, avoiding the ventrolateral portion (Fig. 5). Since no other area in the hypo-

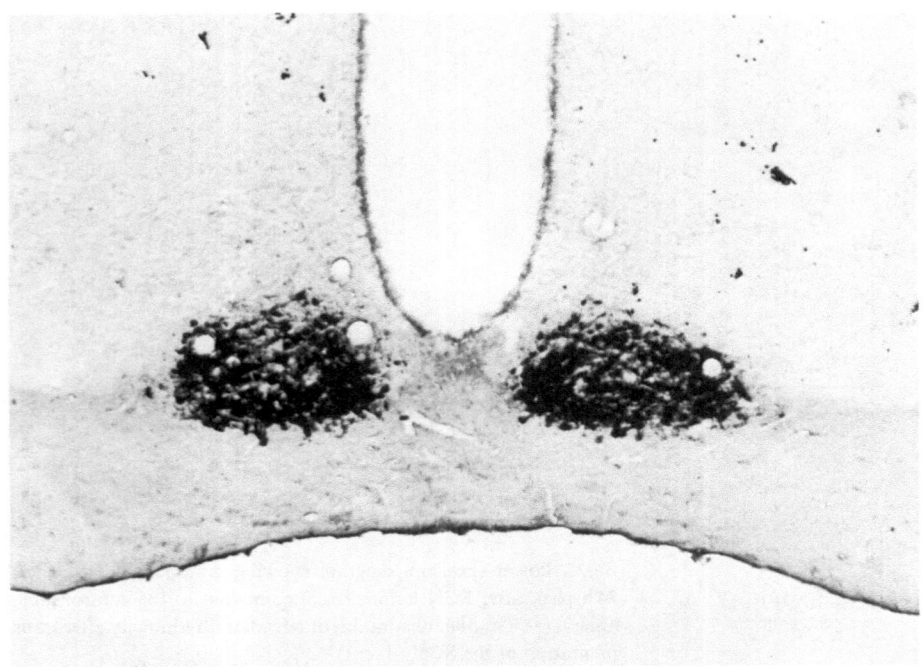

Fig. 4. Location of vasoactive intestinal polypeptide (VIP) containing neurons in the ventrolateral portion of the SCN

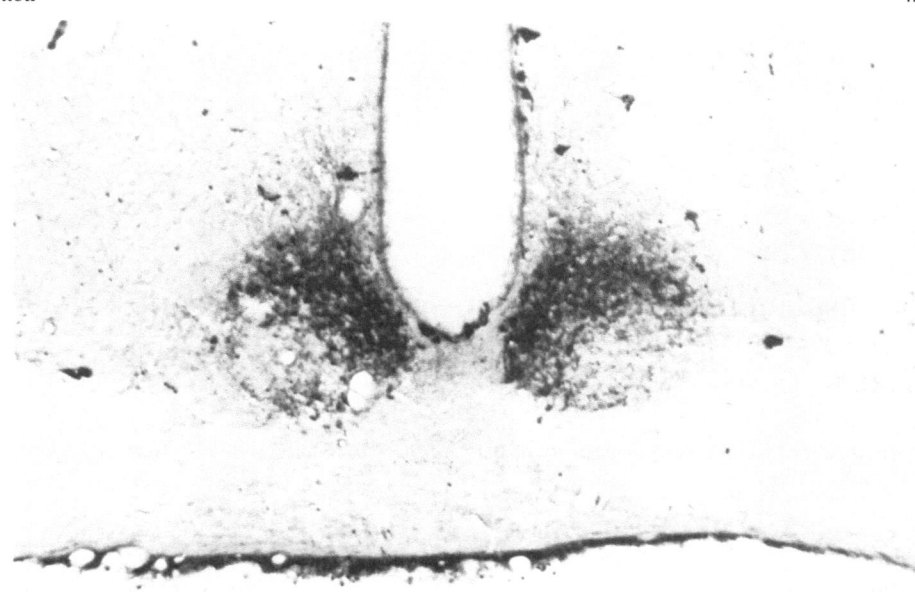

Fig. 5. Location of vasopressin-containing neurons in the dorsomedial portion of the SCN

system includes the posterior hypothalamus. Therefore, the SCN transplanted into the caudal wall of the third ventricle would easily affect the neural activity of the whole brain through a non-specific reticular activating system.

Acknowledgements

Authors wish to thank Miss Yoko Mori for her technical assistance.

References

1. Card JP, Brecha N, Karten HJ, Moore RY (1981) Immunocytochemical localization of vasoactive intestinal polypeptide-containing cells and processes in the suprachiasmatic nucleus of the rat: light and electron microscopic analysis. J Neurosci 1: 1289–1308
2. Ibuka N, Inouye ST, Kawamura H (1977) Analysis of sleep-wakefulness rhythms in male rats after suprachiasmatic nucleus lesions and ocular enucleation. Brain Res 122: 33–47
3. Ibuka N, Kawamura H (1975) Loss of circadian rhythm in sleep-wakefulness cycle in the rat by suprachiasmatic nucleus lesion. Brain Res 96: 76–81
4. Inouye ST, Kawamura H (1979) Persistence of circadian rhythmicity in a mammalian hypothalamic "island" containing the suprachiasmatic nucleus. Proc Natl Acad Sci 76: 5962–5966
5. Inouye ST, Kawamura H (1982) Characteristics of a circadian pacemaker in the suprachiasmatic nucleus. J Comp Physiol 146: 153–160
6. Kawamura H, Inouye ST (1979) Circadian rhythm in a hypothalamic island containing the suprachiasmatic nucleus. In: Suda M, Hayaishi O, Nakagawa H (eds) Biological rhythms and their central mechanism. Biomedical Press, Amsterdam, Elsevier/North Holland, pp 335–341
7. Kawamura H, Nihonmatsu I (1985) The suprachiasmatic nucleus as a circadian rhythm generator: immunocytochemical identification of the suprachiasmatic nucleus within the transplanted hypothalamic tissues. In: Hiroshige T, Honma K (eds) Circadian clocks and Zeitgebers. Hokkaido University Press, Sapporo, pp 55–63
8. Kawamura H, Nihonmatsu I, Saitoh Y, Sato T, Sawaki Y (1984) Transplantation of neonatal suprachiasmatic nuclei into the rat brain. Neurosci Lett [Suppl] 17: 31
9. Richter C (1967) Sleep and activity: their relation to the 24 h clock. In: Kety SS, Evarts EV, Williams LL (eds) Sleep and altered states of consciousness. Proc Res Assoc Nerv Ment Dis, Williams and Wilkins, Baltimore, pp 8–29
10. Sawaki Y, Kawamura H (1983) Transplantation of the suprachiasmatic nucleus in rats subjected to bilateral lesions of the suprachiasmatic nuclei. J Physiol Soc Jpn 45: 462
11. Sawaki Y, Nihonmatsu I, Kawamura H (1984) Transplantation of the neonatal suprachiasmatic nuclei into rats with complete bilateral suprachiasmatic lesions. Neurosci Res 1: 67–72
12. Van den Pol AN, Powley TL (1979) A fine grained anatomical analysis of the role of the rat suprachiasmatic nucleus in circadian rhythms of feeding and drinking. Brain Res 160: 307–326

Correspondence: Dr. Yoshito Saitoh, Department of Neuroscience, Mitsubishi Kasei Institute of Life Sciences, 11, Minamiooya, Machida-shi, Tokyo 194, Japan.

Acta Neurochirurgica, Suppl. 41, 46–50 (1987)

Transplantation to the Brain—a New Therapeutic Principle or Useless Venture?

E.-O. Backlund

Department of Neurosurgery, University of Bergen School of Medicine, Bergen, Norway

Summary

On the basis of remarkable results from transplantation experiments in animals, the author in 1979 suggested autologous transplantation of cathecholamine tissue to the brain as tentative treatment for parkinsonism. Experiences from four patients thus treated are reported comprehensively. Transitory rewarding effects were seen when the transplant was deposited in the putamen. The potential prospects for this principally new therapeutic paradigm are discussed, as well as possible improvements in the method.

Keywords: Transplantation; parkinsonism; stereotaxy; adrenal medulla.

Surgical repair of the CNS must have been a dream for clinical neuroscientists since the first days of neurosurgery and already at the turn of the century, some trials were performed. It is beyond the scope of this article to give a full account of the history of CNS transplantation, but as an example I find it appropriate to give credit to two clear-sighted pioneers. One fundamentally interesting attempt was made by Saltykow[19], who excised pieces of the cortex in rabbits and re-implanted them immediately at the same site, to assess their usefulness as components for restoration of CNS injuries.

Saltykow's design should reasonably give the graft optimal chances to survive. Nevertheless, the results were disappointing, however conclusive: mature CNS neurons cannot stand even a very short and gentle separation from their normal environment. Saltykow's pioneer results were already published in 1905.

Another fundamental study by Hopkins Dunn[5], published in 1917, showed that a certain ability to survive transplantation could be seen if immature CNS neurons were used. Cortex grafts from ten-day-old rats were taken and implanted into cortical cavities in litter mate recipients. It was found that graft survival correlated consistently with the vascularization; those grafts which, by chance, were in contact with the choroid plexus developed a vascular pattern and survived. On the other hand, no morphological contacts between the transplanted cells and the host neurons were established across the glial barrier surrounding the graft. These two conclusions, that very young cells are less vulnerable and that a vascular network is crucial for the vitality of the graft, became important starting points when later projects were designed.

The enormous boom in basic neuroscience due to the recent developments of unique and ingenious laboratory methods, implied dramatically new conditions for experimental transplantation in the CNS. Elegant and specific immunohistochemical methods, sophisticated tissue culturing and the establishment of genetically "tailor-made" strains of research animals are only a few examples to be mentioned.

In many of the current experimental projects, it is not by chance that attention has been directed to the possible treatment of parkinsonism, using transplantation procedures. First, the pathogenetic mechanisms of this disease have been extensively studied and partly understood, making it a challenging domain for clinical application of the results from experimental neurobiology. Secondly, investigations in a couple of specially designed animal models have proved to be useful substitutes for clinical studies. One, now almost a classic, is "the rotating rat", first described by Ungerstedt[23], with specific lesion of the nigro-striatal dopaminergic system, induced by microinjection of 6-hydroxydopamine. The other, more spectacular and newly discovered experimental design uses primates, lesioned with MPTP (1-methyl-4-phenyl,1,2,3,6-

tetrahydropyridine), giving an almost uncomfortable resemblance to human parkinsonism[4]. A lot of information has further been obtained from *in oculo* experiments, when transplants have been studied growing on the sympathectomized iris, in the anterior chamber of the rat's eye (for a review, see[17]).

In 1979, a signal paper appeared in Science by Perlow et al.[18], using the "rotating rat" model. It was proven, that a fetal substantia nigra graft transplanted to the denervated striatum not only survived and proliferated but also induced a return to normal of the lesion-induced motor abnormality. Later, even more spectacular studies showed that the abnormality reappeared when the graft was removed.

The experiments thus proved that immature neurons can survive transport from one site to another within the CNS and even continue to express neuro-humoral and electrophysiological characteristics in the new environment[7, 24]. Surprisingly, this was also shown for adult tissue[16] and across species borders[2]. Further, region-specific trophic factors obviously influenced the grafted tissue in its new environment, sometimes to an incredible degree, by transformation of transplanted cells to a new cell type. Thus, it was shown that AM cells transformed themselves to a neuron-like cell type, with a tendency to fibre production[22]. Such morphological changes mirror a tendency of the transplanted cells not only to be able to adapt themselves to the new environment but also to influence it reciprocally.

From the clinician's point of view, certainly all this experimental evidence is a virtual challenge. For a neurosurgeon familiar with the modern marriage between sophisticated imaging techniques and stereotactic surgery, it seems reasonable that the animal experiments might be scaled up to clinical dimensions.

The laboratory results became the impetus for the author to suggest the initiation of a clinical study. An application for approval of a patient project including ten cases was submitted to the local ethical committee at the Karolinska Hospital, Stockholm, Sweden. After some further animal experiments, it was approved in 1981. In March 1982, the first transplantation to the human brain was performed[1].

One of the ideas behind this clinical trial was to explore the possibilities of designing an operation also for rigid and hypokinetic Parkinson patients. To a neurosurgeon, it is only too well-known that thalamotomy, the established operation in parkinsonism, is rewarding in patients with tremor whereas one has to trust in medication for rigid and hypokinetic patients.

When selecting patients for our project, this was taken into account.

Using an autologous approach, we wanted decisively to combine the immunological advantages of grafting from the patient himself with the escape from ethical objections. Grafts from the cervical sympathetic chain of the patient was discussed, but we fixed our choice on the adrenal medulla (AM), which in the animal experiments had shown a certain ability to express neuronal characteristics when transplanted to the brain.

Our attention was further directed to the possible use of trophic factors together with the transplant. Results from the anterior eye chamber experiments indicated an enhanced proliferation of AM grafts after the administration of nerve growth factor (NGF)[21]. It was only after the first clinical trials, however, that we fully appreciated these results. This will be taken into account in the next series.

The clinical project will now be described briefly, with first of all, some comments on the target selection. In the animals, the transplants most often were introduced into the central area of the striatum. In rodents, this complex is dominated by the head of the caudate nucleus, whereas the "putaminal" part, if any, is insignificant. Thus, the caudate was arbitrarily selected as recipient area in our first two patients. However, later and significant studies by another Swedish group in post-mortem wet-brain analysis indicated the putamen of parkinsonian patients to be more depleted of dopamine than the caudate[15]. Interestingly, the dopamine concentration in putamen was also higher than that of the caudate in a matching group of normals. Also positron emission studies, using dopamine receptor ligands, show a higher receptor density in the putamen, as compared to the caudate, when normal brains are studied[6].

For the operations, patients of akinesia-rigidity type were selected and L-dopa must have been proven effective. In the first two patients, the degree of disability was assessed by standard neurological examination and a self-scoring system. In these two, the aim was also to keep them drug-free as long as possible. In the first, the L-dopa medication was stopped ten days before operation and continued for another ten days after the operation. The second patient was in a more serious condition but could stand a four-day drug-free period post-operatively. Patients 3 and 4 were more thoroughly assessed, with ambitious motor performance tests, sensory- and movement-evoked potentials, quantitative EEG recordings, measurements of regional blood flow and monoamines in the lumbar CSF, positron emission tomography using a dopamine receptor ligand and CT scans. They had two drug-free periods, the first as a test during four days in the pre-operative period and a second four-day period, starting on the day of operation.

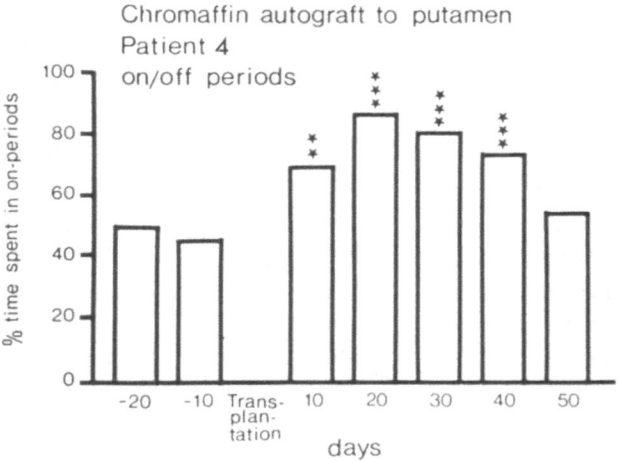

Fig. 2. Patient 4: Before the operation, the patient spent approximately 50% of his time in on-periods. During the first 50 post-operative days, with a maximum in the third week, the time spent in on-periods was significantly increased (**: $p < 0.01$; ***: $p < 0.001$), compared to pre-operative values

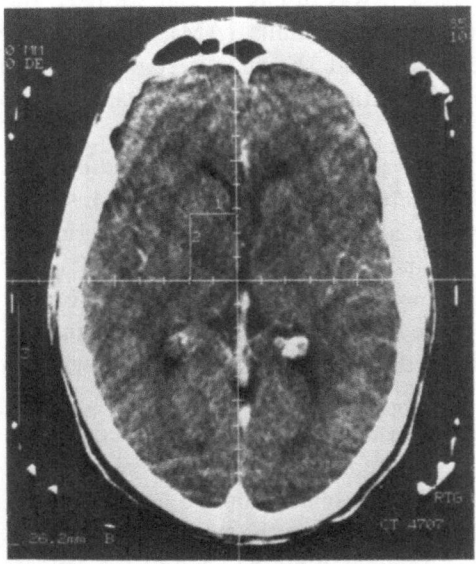

Fig. 1. Stereotactic CT during the operation. The lateral border of the striatum is well seen. The stereotactic coordinates for a target in the putamen are indicated on the superimposed grid

Using CT-guided stereotactic technique[12], one (patient 1) or two (patients 2–4) 16–19 mm columns of 1 mm autologous adrenal medulla fragments were introduced into the central area of the head of the caudate nucleus (patients 1–2) and putamen (patients 3–4, Fig. 1), respectively. The treatment protocol and the surgical procedure have been described elsewhere[1, 13].

In the first two patients, a discrete amelioration of the rigidity was found for some few days after the operation, but no lateralization of this effect. Soon, however, they regained their pre-operative state and they had to use their previous medication again. In the other two, the effect was more pronounced, especially in patient 4, who had a significant improvement in this on-off periodicity (Fig. 2). But at the six-month follow-up, patients 3 and 4 also were in the same state as pre-operatively[13].

Now, was this a venture in vain or has possibly a radically new therapeutic paradigm been introduced? Future studies will give a more definite answer, but new principles for the treatment of parkinsonism have doubtlessly been outlined. As the present project con-

cerns, a prudent approach to the above-mentioned question would be to look for new options in the treatment protocol and for improvements of the technique. The experiences hitherto give reasons for some new routes to follow, for instance:

1. The optimal recipient area in the patient's brain (within the striatum?, elsewhere?) may be sought for by explorative microinjection of dopamine or other substances.

A sophisticated technique for this purpose would be the microdialysis, designed by Ungerstedt[25]. Using this during the operation also gives the opportunity to harvest samples from the *milieu interieur* of the target area.

In the animals, functional effects were obtained when the graft was placed centrally in the caudate, whereas the same procedure performed in more peripheral parts of the striatum was much less effective[9]. A possible spatial, "somatotopic" organization of functional subunits within the striatum of man is thus an obvious field for further exploration.

2. As mentioned above, the use of trophic factors, most probably on a long-term basis using an implantable pump system, should also be explored in future cases. The fairly short-lasting and limited response in our four patients indicate a limited survival of the graft *or* its declining ability to produce catecholamines. This may be counteracted with the use of NGF, other trophic factors or gangliosides, *e.g.*

3. In the animal experiments, transplanted cell suspension seems to be superior to solid grafts, albeit in

small fragments. The human AM tissue is very tough and has so far baffled endeavours to become dissociated, but new techniques are under consideration.

4. From many aspects, an operation without general anaesthesia would be advantageous, as this would make possible a detailed study of the patient's peroperative and acute response to the implantation, in neurological terms. If AM grafts are used, the adrenalectomy can then be performed at a separate procedure, preceeding the transplantation, allowing the graft to be preserved/prepared during a number of days. Further studies on the survival ability of human cells in culture would imply new preservation techniques, with the aim of enhancing the quality of the transplant.

5. The future may also offer new alternatives in transplants derived from cell lines. The so-called PC 12 cells (cloned and "domesticated" from rat phaeochromocytoma) have been used in a number of animal studies, with partly exciting results[8]. In a clinical project a potential alternative would also be a human cell line, manipulated to serve the purpose of catecholamine production, by genetic engineering. Further, as the NGF gene is known, transfected human cells producing NGF would be possible to use as a co-graft.

Cathecholamine-producing brain tissue (substantia nigra) from aborted human fetuses has been suggested as a tempting alternative, but serious ethical implications have to be analysed more thoroughly before this route may be chosen[14]. Nonetheless, approval from both national and local ethical committees in Sweden have been given. This is something of a paradox, as we have experienced a certain ambivalence, both ethically and scientifically. An unbiased decision in this crucial question is especially difficult to make, as intriguing results from on-going animal experiments show, that the human fetal neuroblasts have a remarkable vitality and plasticity in the host brain, and counteract 6-hydroxydopamine-induced motor dysfunction effectively[3, 10, 20]. These experiments have included immuno-suppression, using cyclosporin A.

Such improvements and further development of the clinical experimental design, combined with advances in basic and clinical neurosciences, may imply progress to the transplantation issue. Still, we are far from the goal of new therapeutic routines for cruel degenerating diseases in the CNS, but the experiences presented here indicate, at least, that future neurosurgeons will meet radically new and exciting challenges.

References

1. Backlund EO, Granberg PO, Hamberger B et al (1985) Transplantation of adrenal medullary tissue to striatum in parkinsonism: first clinical trials. J Neurosurg 62: 169–173
2. Björklund A, Stenevi U, Dunnett SB et al (1982) Cross-species neural grafting in a rat model of Parkinson's disease. Nature 298: 652–654
3. Brundin P, Nilsson OG, Strecker RE et al (1987) Behavioural effects of human fetal dopamine neurons grafted in a rat model of Parkinson's disease. Exp Brain Res (in press)
4. Burns RS, Chieuh CC, Markey SP et al (1983) A primate model of parkinsonism: Selective destruction of dopaminergic neurons in the pars compacta of the substantia nigra by N-methyl-4-phenyl-1,2,3,6-tetrahydropyridine. Proc Natl Acad Sci USA 80: 4546–4550
5. Dunn EH (1917) Primary and secondary findings in a series of attempts to transplant cerebral cortex in the albino rat. J Comp Neurol 27: 565–582
6. Farde L, Hall H, Ehrin E et al (1985) Dopamine-D2 receptor density determined in the living human brain by saturation curves. Acta Neurol Scand 72: 234
7. Freed WJ, Perlow MJ, Karoum F et al (1980) Restoration of dopaminergic function by grafting of fetal rat substantia nigra to the caudate nucleus: Long-term behavioural, biochemical and histochemical studies. Ann Neurol 8: 510–519
8. Hefti F, Hartikka J, Schlumpf M (1985) Implantation of PC 12 cells into the corpus striatum of rats with lesions of the dopaminergic migrostriatal neurons. Brain Res 348: 283–288
9. Herrera-Marschitz M, Strömberg I, Olsson D et al (1984) Adrenal medullary implants in the dopamine-denervated rat striatum. II. Acute behaviour as a function of graft amount and location and its modulation by neuroleptics. Brain Res 297: 53–61
10. Kamo H, Kim SU, McGeer PL et al (1986) Functional recovery in a rat model of Parkinson's disease following transplantation of cultured human sympathethic neurons. Brain Res 397: 372–376
11. Langston JW, Ballard P, Tetrud JW et al (1983) Chronic parkinsonism in humans due to a product of meperidine-analog synthesis. Science 219: 979–980
12. Leksell L, Jernberg B (1980) Stereotaxis and tomography. A technical note. Acta Neurochir (Wien) 52: 1–7
13. Lindvall O, Backlund EO, Farde L et al (1987) Transplantation in Parkinson's disease: two cases of adrenal medullary grafts to putamen. Ann Neurol 22: 457–468
14. Murphy PJ (1984) Moral perspectives in the use of embryonic cell transplantation for correction of nervous system disorders. Appl Neurophysiol 47: 65–68
15. Nyberg P, Nordberg A, Wester P et al (1983) Dopaminergic deficiency is more pronounced in putamen than in nucleus caudatus in Parkinson's disease. Neurochem Pathol 1: 193–202
16. Olson L (1970) Fluorescence histochemical evidence for axonal growth and secretion from transplanted adrenal medullary tissue. Histochemie 22: 1–7
17. Olson L, Backlund EO, Freed W et al (1985) Transplantation of monoamine-producing cell systems *in oculo* and intracranially: experiments is search of a treatment for Parkinson's disease. In: Nottebohm F (ed) Hope for a new neurology. Ann NY Acad Sci 457: 105–126
18. Perlow MJ, Freed WJ, Hoffer BJ et al (1979) Brain grafts reduce motor abnormalities produced by destruction of nigrostriatal dopamine system. Science 204: 643–647

19. Saltykow S (1905) Versuche über Gehirnreplantation, zugleich ein Beitrag zur Kentniss reaktiver Vorgänge an den zelligen Gehirnelementen. Arch Psychiat 40: 329–388

20. Strömberg I, Bygdeman M, Goldstein M et al (1986) Human fetal substantia nigra grafted to the dopamine-denervated striatum of immuno-suppressed rats: Evidence for functional reinnervation. Neurosci Lett 71: 271–276

21. Strömberg I, Ebendal T, Seiger Å et al (1985) Nerve fiber production by intraocular adrenal medullary grafts: Stimulation by nerve growth factor or sympathetic denervation of the host iris. Cell Tissue Res 241: 241–249

22. Strömberg I, Herrera-Marschitz M, Ungerstedt U et al (1985) Chronic implants of chromaffin tissue into the dopamine-denervated striatum. Effects of NGF on graft survival, fiber growth and rotational behaviour. Exp Brain Res 60: 335–349

23. Ungerstedt U, Arbuthnott G (1970) Quantitative recording of rotational behaviour in rats after 6-hydroxydopamine lesions of the nigro-striatal dopamine system. Brain Res 24: 485–493

24. Wuerthele SM, Olson L, Freed W et al (1984) Electrophysiology of substantia nigra transplants. In: Usdin E, Carlsson A, Dahlström A et al (eds) Catecholamines. Part B. Neuropharmacology and central nervous system: theoretical aspects. Alan R Liss, New York, pp 333–341

25. Zetterström T, Sharp T, Marsden CA et al (1983) *In vivo* measurement of dopamine and its metabolites by intracerebral dialysis: Changes after D-amphetamine. J Neurochem 41: 1769–1773

Correspondence: Prof. E.-O. Backlund, M.D., Ph.D., Department of Neurosurgery, Haukeland Hospital, N-5016 Bergen, Norway.

Acta Neurochirurgica, Suppl. 41, 51–67 (1987)
© by Springer-Verlag 1987

Self-organizing Neural Network Models for Visual Pattern Recognition

K. Fukushima

NHK Science and Technical Research Laboratories, Kinuta, Setagaya, Tokyo, Japan

Summary

Two neural network models for visual pattern recognition are discussed. The first model, called a "neocognitron", is a hierarchical multilayered network which has only afferent synaptic connections. It can acquire the ability to recognize patterns by "learning-without-a-teacher": the repeated presentation of a set of training patterns is sufficient, and no information about the categories of the patterns is necessary. The cells of the highest stage eventually become "gnostic cells", whose response shows the final result of the pattern-recognition of the network. Pattern recognition is performed on the basis of similarity in shape between patterns, and is not affected by deformation, nor by changes in size, nor by shifts in the position of the stimulus pattern.

The second model has not only afferent but also efferent synaptic connections, and is endowed with the function of selective attention. The afferent and the efferent signals interact with each other in the hierarchical network: the efferent signals, that is, the signals for selective attention, have a facilitating effect on the afferent signals, and at the same time, the afferent signals gate efferent signal flow. When a complex figure, consisting of two patterns or more, is presented to the model, it is segmented into individual patterns, and each pattern is recognized separately. Even if one of the patterns to which the models is paying selective attention is affected by noise or defects, the model can "recall" the complete pattern from which the noise has been eliminated and the defects corrected.

Keywords: Neural network model; self-organization; visual pattern recognition; selective attention; associative recall.

1. Introduction

There are a great number of neural cells in the brain, more than ten billion in the case of a human being. These cells have complicated interconnections and constitute a network on a huge scale. Hence, it is difficult to uncover the neural mechanism of the higher functions of the brain.

In the conventional neurophysiological approach, we use, for example, a micro-electrode to record the response of these cells. The recording can be made, however, only from one cell at a time, or, at most, only from a few cells simultaneously. Although we can thus obtain a vast amount of fragmentary knowledge, it is not easy to understand the mechanism of the network as a whole in this way.

Hence, we tackle the problem from a different approach: we are studying how to interconnect neurons in order to synthesize a brain model, that is, a network which has the same function and ability as the brain. When synthesizing a model, we try to follow physiological evidence as faithfully as possible. For parts which are not clear yet, however, we build up a hypothesis, and synthesize a model following the hypothesis. Then we analyse or simulate the behaviour of the model, and compare it with that of the brain. If any discrepancy in the behaviour is found between the model and the brain, we change the initial hypothesis, and modify the model following a new hypothesis. We then test the behaviour of the model again. We repeat this procedure again and again until the model behaves in the same way as the brain. Although it is still necessary to verify the validity of the model by physiological experiment, it is very probable that the brain works with the same mechanism as the model, because they both respond in the same way. Hence, modelling neural networks is a promising approach to uncovering the mechanism of the brain.

In this paper, we concentrate our discussion on the mechanism of pattern recognition and of self-organization, and propose two neural network models. In the first model, we consider only afferent signals in the brain. The second model has not only afferent but also efferent synaptic connections between the cells in the network, and is endowed with a function of selective attention.

The first model[3, 5], called a "neocognitron", is a hierarchical multilayered network consisting of a cascade of many layers of simplified neural cells. It has afferent synaptic connections between cells in adjoining cell layers. It can acquire the ability to recognize patterns by a so-called "learning-without-a-teacher" process. During the process of learning, synaptic connections between the cells, which are modifiable, grow gradually in accordance with the stimuli given to the network. The repeated presentation of a set of training patterns is sufficient for the self-organization of the network, and it is not necessary to give any information about the categories in which these patterns should be classified. At the highest stage of the network, only one cell eventually responds to each training pattern. Other cells respond to other training patterns. Thus, the cells of the highest stage may be called "gnostic cells", whose response shows the final result of the pattern-recognition of the network. After finishing the process of learning, pattern recognition is performed only on the basis of similarity in shape between patterns, and is not affected by deformation, nor by changes in size, nor by shifts in the position of the stimulus patterns.

The second model[7] also consists of a hierarchical neural network, but it has efferent as well as afferent synaptic connections between cells. The afferent and the efferent signals interact with each other in the network: the efferent signals, that is, the signals for selective attention, have a facilitating effect on the afferent ones, and, at the same time, the afferent signals gate efferent signal flow. When a complex figure consisting of two patterns or more is presented to the model, it is segmented into individual patterns, and each pattern is recognized separately. Even if one of the patterns to which the model is paying selective attention is affected by noise or defects, the model can recall the complete pattern from which the noise has been eliminated and the defects corrected. It is not necessary for perfect recall that the stimulus pattern should be identical in shape to the training pattern. Even though the pattern is distorted in shape or changed in size, it can be correctly recognized and the missing portions restored.

Various models for pattern-recognition or associative memory have been reported to be able to recognize patterns or to recall complete patterns from imperfect ones[6, 11, 14]. Most of the earlier models, however, do not work well unless the stimulus pattern is identical in size, shape and even position to a training pattern. In contrast to such earlier models, the models discussed in this paper have the ability of pattern-

recognition and perfect autoassociative recall, even for deformed patterns and regardless of their position.

2. Neocognitron [3, 5]

2.1. The Structure and Behaviour of the Network

In the visual cortex, neurons are found to respond selectively to local features of a visual pattern, such as lines and edges in particular orientations[8, 9]. In the area higher than the visual cortex, it has been found that cells exist which respond selectively to certain figures like circles, triangles or squares[15], or even to a human face[1]. Accordingly, the visual system seems to have a hierarchical structure, in which simple features are first extracted from a stimulus pattern, and then integrated into more complicated ones. In this hierarchy, a cell in a higher stage generally has a larger receptive field, and is more insensitive to the position of the stimulus. This kind of physiological evidence suggested a network structure for the neocognitron.

The neocognitron is a multilayered network consisting of a cascade of many layers of neural cells. The cells are of the anologue type; that is, their inputs and output take non-negative analogue values, corresponding to the instantaneous firing-frequencies of biological neurons. Fig. 1 shows a typical example of the cells employed in the network.

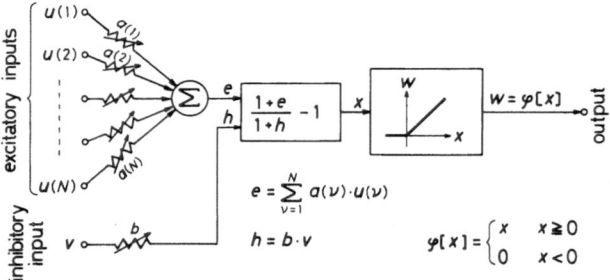

Fig. 1. Input-to-output characteristics of a u_S-cell: a typical example of the cells employed in the neocognitron

The hierarchical structure of the network is illustrated in Fig. 2. There are afferent synapses between cells in adjoining cell-layers. Let the initial stage of the hierarchical network be called U_0, the l-th stage U_l, and the highest stage U_L. Incidentally, Fig. 2 shows the case of L = 3. At each stage, various kinds of cells, such as u_S, u_C, u_{SV}, are severally arranged in two-dimensional arrays. Notation u_{Sl}, for example, is used to denote a u_S-cell in the l-th stage. We also use notation such as U_{Sl} to denote the layer of u_{Sl}-cells.

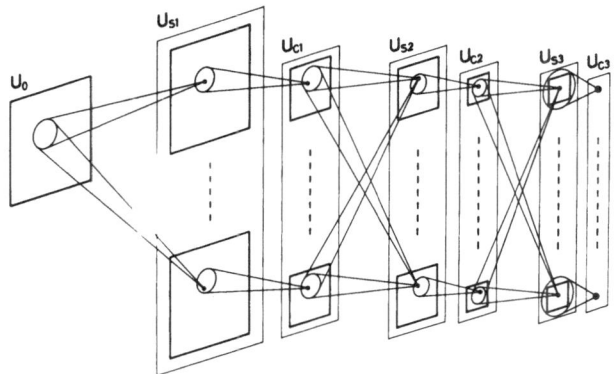

Fig. 2. Hierarchical network structure of the neocognitron

The initial stage U_0 is the input layer of the network, and consists of a two-dimensional array of receptor cells u_0. (This does not necessarily mean that u_0-cells are photoreceptors in the retina. They might correspond to cells in a higher stage, say, to the lateral geniculate cells.) Each of the succeeding stages has a layer of u_S-cells followed by a layer of u_C-cells. u_S-cells are feature-extracting cells and somewhat resemble simple cells in the visual cortex, while u_C-cells resemble complex cells. u_S-cells or u_C-cells in a layer are divided into subgroups according to the optimum stimulus features of their receptive field. Since the cells in each subgroup are arranged in a two-dimensional array, we call the subgroup a "cell-plane". In Fig. 2, each quadrangle drawn with heavy lines represents a cell-plane, and each vertically elongated quadrangle drawn with thin lines, in which cell-planes are enclosed, represents a layer of u_S-cells or u_C-cells. As illustrated in Fig. 3, all the cells in a cell-plane receive synaptic connections of the same spatial distribution, and only the positions of the presynaptic cells are shifted in parallel from cell to cell. Although cells usually exist in numbers, only one cell is drawn in each cell-plane in Fig. 2. Each ellipse in the

figure represent the area from which a cell receives synaptic connections.

The density of cells in a layer is designed to decrease with the order of the stage, because the cells in the higher stages usually have larger receptive fields. Thus, in the highest stage, only one u_C-cell exists in each cell-plane.

Let us consider the mechanism of pattern recognition in the network in more detail. Synaptic connections converging to feature-extracting cells u_S have plasticity and can be modified during a learning (or training) process. After finishing the learning, which will be discussed later, u_S-cells, with the aid of the inhibitory cells u_{SV}, can extract features from the stimulus pattern. In other words, a u_S-cell is activated only when a particular feature is presented at a certain position in the input layer. The features which the u_S-cells extract are automatically chosen by the network itself during the learning process. Generally speaking, in the lower stages, local features, such as a line at a particular orientation, are extracted. In the higher stages, more global features, such as a part of a training pattern, are extracted.

Fig. 4 is a detailed diagram illustrating the synaptic connections converging to a feature-extracting cell u_S. Through the excitatory synapses coming directly from u_C-cells of the preceding stage, the u_S-cell receives signals indicating the existence of the relevant feature to be extracted. The inhibitory cell u_{SV} is always responding with the average intensity of the output of the u_C-cells presynaptic to be u_S-cell. If an irrelevant feature is presented, the inhibitory signal from the u_{SV}-cell becomes stronger than the direct excitatory signals from u_C-cells, and the response of the u_S-cell is suppressed. Hence, the inhibitory u_{SV}-cell can be said to be watching for the existence of irrelevant features. Thus inhibitory cells u_{SV} play an important role in

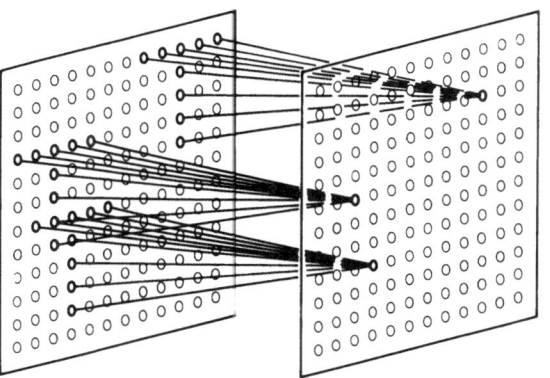

Fig. 3. Illustration showing the spatial arrangement of the synaptic connections converging to single cells of a cell-plane

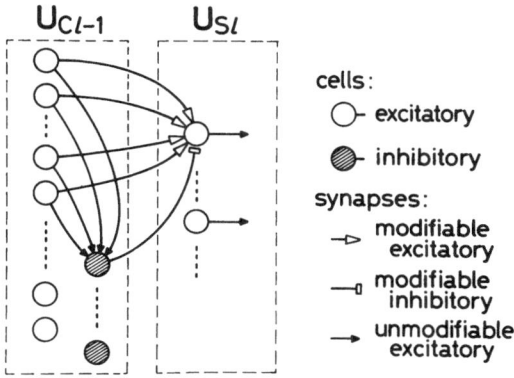

Fig. 4. Synaptic connections converging to a feature-extracting cell u_S

endowing the feature-extracting cells u_S with the ability to differentiate irrelevant features, and in increasing the selectivity of feature extraction.

In each stage, the layer of u_C-cells follows the layer of u_S-cells. Synaptic connections from u_S-cells to u_C-cells are fixed and unmodifiable. Each u_C-cell receives signals from a group of u_S-cells, all of which extract the same feature but from slightly different positions. The u_C-cell is activated if at least one of these u_S-cells is active. Even if the feature is presented at a slightly different position, the same u_C-cell keeps responding. Hence, the u_C-cell's response is less sensitive to shifts in position of the stimulus pattern.

Thus, in the whole network, in which layers of u_S-cells and u_C-cells are arranged alternately, the process of feature extraction by u_S-cells and toleration of positional shift by u_C-cells are repeated. During this process, local features extracted in a lower stage are gradually integrated into more global features. Fig. 5 illustrates this situation schematically. Finally, each u_C-cell of the highest stage integrates all the information of the stimulus pattern, and responds only to one specific pattern. In other words, in the highest state, only one u_C-cell, corresponding to the category of the stimulus

pattern, is activated. Other cells respond to patterns of other category. Thus, the u_C-cells of the highest stage may be called "gnostic cells", and their response is the final result of the pattern-recognition of the network.

The operation of tolerating positional error a little at a time at each stage, rather than all in one step, plays an important role in endowing the network with an ability to recognize even distorted patterns. Since errors in the relative position of local features are tolerated in the process of extracting and integrating features, the same u_C-cell responds in the highest stage, even if the stimulus pattern is deformed or changed in size or shifted in position. In other words, the neocognitron recognizes the "shape" of the pattern independent of its size and position.

2.2. Self-organization of the Network

The neocognitron acquires the ability to recognize patterns by so-called "learning-without-a-teacher" process or "unsupervised learning". The repeated presentation of a set of training patterns is sufficient for the self-organization of the network, and it is not necessary to give any information about the categories in which these patterns should be classified. The neocognitron by itself acquires the ability to classify and recognize these patterns correctly, on the basis of similarity in shape.

Self-organization of the neocognitron is performed on two hypotheses (rules). The first has been used for the self-organization of the "cognitron"[4] proposed earlier by the author. Specifically, the hypothesis is as follows:

The modifiable synaptic connection between two cells is reinforced if and only if the following two conditions are simultaneously satisfied:

(1) The postsynaptic cell is responding the most strongly among the cells in its vicinity.

(2) The presynaptic cell is also responding.

This hypothesis can also be expressed as follows:

Among the postsynaptic cells situated in a certain small area, only the one which is responding most strongly has its input synpases reinforced. The amount of reinforcement of each input synapse to this maximum-output cell is proportional to the intensity of the response of the relevant presynaptic cell.

Hence, the input synapses to the maximum-output cell grow so as to work as a "template" which exactly matches the spatial distribution of the response of the presynaptic cells. Thus, the maximum-output cell comes to acquire the ability to extract the feature of the

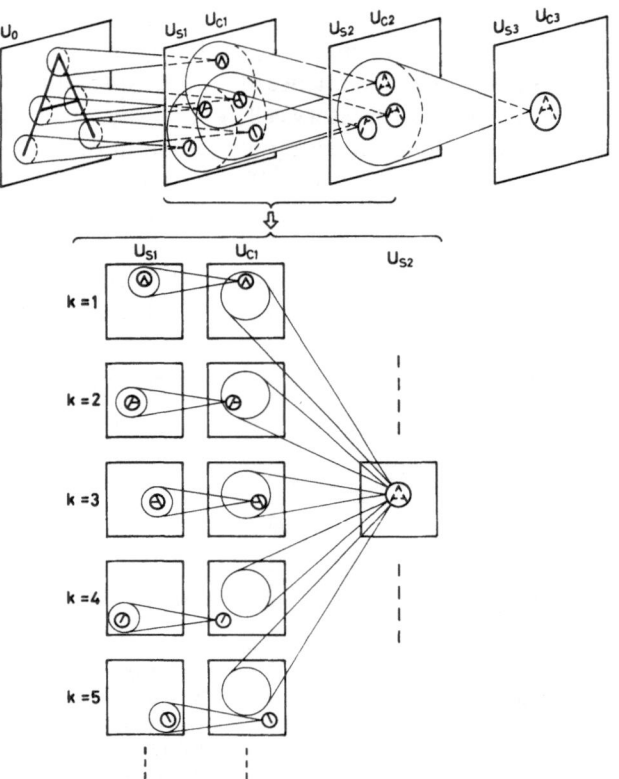

Fig. 5. Illustration of the processes of pattern recognition in the neocognitron

stimulus which has been presented during the training period.

It is assumed that this hypothesis holds not only for excitatory synapses but also for inhibitory synapses. As has been discussed in connection with Fig. 4, reinforcement of inhibitory synapses endows the cell with the ability to differentiate irrelevant stimuli from the relevant one, and increases the selectivity in feature extraction.

According to this hypothesis, among the cells in a certain small area, only one cell which happens to yield the maximum output is selected to have its input synapses reinforced. Because of the "winner-takes-all" nature of this hypothesis, the duplicated formation of cells which extract the same feature does not occur, and the formation of a redundant network can be prevented. This situation resembles, so to speak, "elite education". Only the one cell which gives the best response to a training stimulus is selected, and only that cell is reinfored so as to respond more appropriately to the stimulus.

With this hypothesis, the neural network also develops a self-repairing function. If a cell which has been stronlgy responding to a stimulus is damaged and ceases to respond, another cell, which happens to respond more strongly than other cells, starts to grow and substitute for the damaged cell. Incidentally, the growth of a second cell has been prevented until then, because of the larger response of the first cell.

The second hypothesis introduced for the self-organization of the neocognitron is that the maximum-output cell not only grows, but also controls the growth of neighbouring cells. In other words, the maximum-output cell works, so to speak, like a seed in crystal growth, and neighbouring cells have their input synapses reinforced in the same way as the "seed cell". The process of selecting seed cells in the neocognitron will be discussed below in more detail.

Here, we define a term "hypercolumn" in an extended manner from the definition by Hubel and Wiesel[10]: a hypercolumn is defined here as a group of cells in a layer same position. In other words, each hypercolumn contains all kinds of feature-extracting cells in it, but these cells have receptive fields at approximately the same position. Incidentally, if we rearrange the cell-planes of a layer and stack them in a manner shown in Fig. 6, the cells of a hypercolumn constitutes a columnar structure. Each hypercolumn contains cells from all the cell-planes.

Now, let us assume that a training stimulus is presented to the network. From each hypercolumn, the

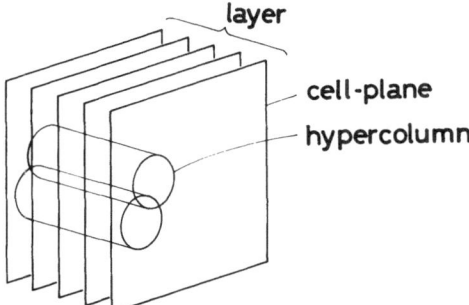

Fig. 6. Relation between cell-planes and hypercolumns within a layer

cell which happens to respond the most strongly is chosen as a candidate for seed cells. When two candidates or more appear in one and the same cell-plane, only the one whose response is the greatest is selected as the seed cell of that cell-plane. When only one candidate appears in a cell-plane, the candidate automatically becomes the seed cell of that cell-plane.

Thus, at most one seed cell is selected from each cell-plane of u_S-cells at a time. Usually, a different cell becomes a seed cell when a different stimulus is given. When a seed cell is selected from a cell-plane, all the other u_S-cells in the cell-plane grow so as to have input synapses of the same spatial distribution as the seed cell. As the result, all the cells in a cell-plane grow to receive synaptic connections of the identical spatial distribution where only the positions of the presynaptic cells are shifted in parallel from cell to cell, as illustrated in Fig. 3. Hence, all the cells in the cell-plane come to respond selectively to a particular stimulus, and differences between these cells arise only from differences in the position of their receptive fields.

If efficiencies of all the modifiable synapses are zero at the initial state before learning, self-organization of the network cannot start, because no cell can respond to the training stimuli and maximum-output cells (or seed cells) cannot be selected. In the neocognitron, it is assumed that all the modifiable excitatory synapses unconditionally get a very small efficiency only when self-organization is going to start. In other words, each u_S-cell temporarily has very weak and diffused excitatory input connections only at the initial period of the self-organization. Once a reinforcement of the input synapses begins, these weak and diffused initial connections are assumed to disappear. This assumption coincides with the anatomical evidence that an enormous number of axon branches appear at the initial period of nerve growth, and that most of them disappear later. Incidentally, if the period of generation of these temporary weak diffused connections is de-

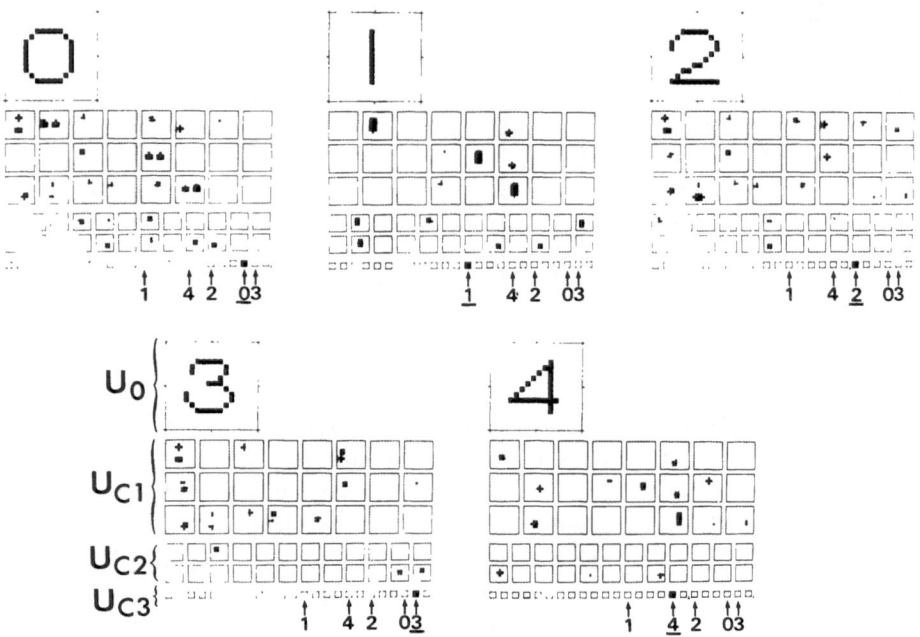

Fig. 7. Response of the u_C-cells in the neocognitron to five stimulus patterns which were used for learning

layed a little for the cells of higher stages, self-organization of the network can be performed efficiently. Specifically, it is desirable to delay it till the growth of the cells of the preceding stage has been settled.

2.3. Computer Simulation

We will now study the behaviour of the neocognitron by computer simulation. The scale of the simulated network is L = 3*.

During the learning period, five training patterns, "0", "1", "2", "3" and "4" shown at the top of each picture in Fig. 7., were repeatedly presented to the network, and self-organization occurred by learning-without-a-teacher. Although the position of the training patterns was shifted randomly every time, we have almost the same results as when the training patterns are always presented in the same position. As a matter of fact, self-organization is generally easier if the position of the patterns is stationary rather than shifted at random. Thus, the experiment under the more difficult conditions is shown here.

After repeated presentation of the five training patterns, the neocognitron gradually acquired by itself the ability to classify the patterns according to differences in their shape. In this simulation, each of the five training patterns was presented 20 times: by that time, self-organization of the network was almost complete.

Fig. 7 shows how the individual cells came to respond to the five training patterns. In each picture, only the response of the u_C-cells in the network, namely, layers U_0, U_{C1}, U_{C2}, U_{C3} are displayed arranged vertically in order. At the highest stage U_{C3}, which is displayed at the bottom of each picture, it can be seen that a different cell responds to a different stimulus. This means that the cells in this layer actually work as "gnostic cells", and that the neocognitron recognizes the five patterns correctly.

Fig. 8 shows how individual cells respond to deformed versions of pattern "2". The response of the cells of intermediate layers, especially the ones near the input layer, varies with the deformation of the stimulus pattern. However, the higher the layer is, the smaller is the variation in the response of the cells in it. Thus, the cells in the highest stage are not affected at all by a deformation of the stimulus pattern; that is, all of the deformed patterns elicited the same response from the cells of layer U_{C3}. Even a deformed pattern which has not been presented to the neocognitron during the learning period is recognized correctly.

---

* The number of cell-planes K_l is 24 for each of the layers U_{S1}–U_{C3}. The numbers of cells in each layer are: 16×16 u_0-cells in U_0; $16 \times 16 \times 24$ u_S-cells, 16×16 u_{SV}-cells, $10 \times 10 \times 24$ u_C-cells in U_1; $8 \times 8 \times 24$ u_S-cells, 8×8 u_{SV}-cells, $6 \times 6 \times 24$ u_C-cells in U_2; $2 \times 2 \times 24$ u_S-cells, 2×2 u_{SV}-cells, and $1 \times 1 \times 24$ u_C-cells in U_3.

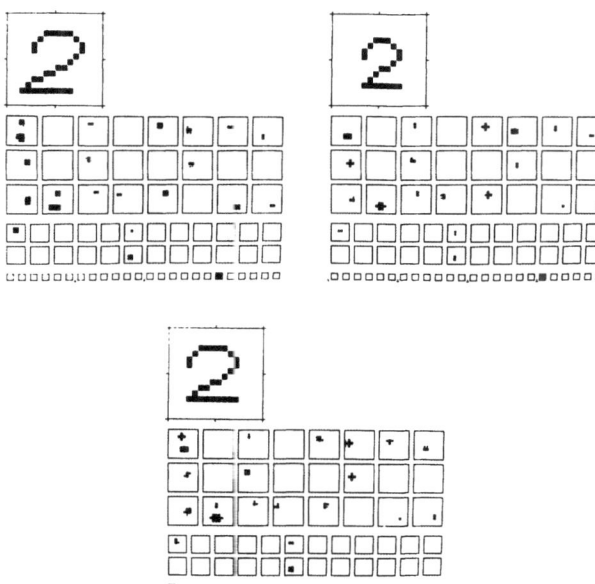

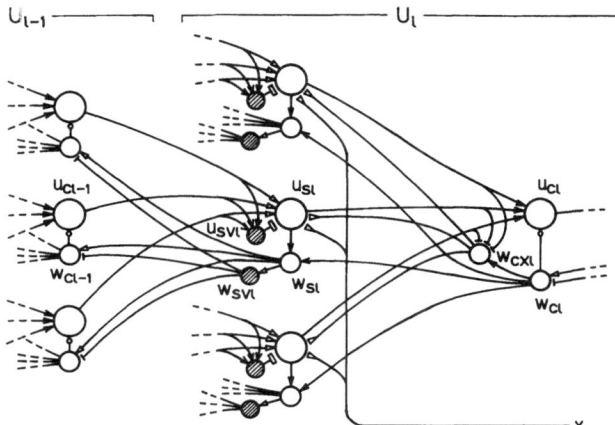

Fig. 10. Detailed diagram illustrating spatial interconnections between neighbouring cells

Fig. 8. Response of the u_C-cells in the neocognitron to deformed patterns

3. Model for Selective Attention[7]

3.1. An Outline of the Structure of the Network

In the model for selective attention, we consider efferent as well as afferent flow of signals in the brain. The model is a hierarchical multilayered network like the neocognitron, but it has not only afferent but also efferent connections between cells.

The hierarchical structure of the network is illustrated in Fig. 9. At each stage, various kinds of cells, such as u_S, u_C, u_{SV}, w_S, w_C, w_{SV} and w_{CX}, are serverally arranged in two-dimensional arrays. The mark ○ in the figure represents a cell*. Although cells of each kind actually exist in numbers, only one cell is drawn in each stage. Between these cells, there are synaptic connections denoted by single lines and double lines in the figure. A single line indicates that there are one-to-one connections between the two groups of cells, and a

double line indicates that there are coverging or diverging connections between them. A detailed diagram, illustrating spatial interconnections between neighbouring cells, is shown in Fig. 10.

Just as in the neocognitron, the density of cells in a layer is designed to decrease with the order of the stage, because the cells in higher stages usually have larger receptive fields. Incidentally, the initial stage U_0 of the network contains two kinds of cells, u_{C0}-cells and w_{C0}-cells. The u_{C0}-cells corresponds to the u_0-cells in the neocognitron.

If we consider the afferent synaptic connections in the network only, and neglect the efferent ones, the model is almost the same as the neocognitron. It can acquire the ability to recognize patterns by learning-without-a-teacher process.

Information about the stimulus given to the input layer of the network is processed at every cell-layer and

---

* In our model, global information flow in the network is mainly taken into account, and the problem of one-to-one correspondence between cells of the model and of the brain has been ignored. Thus, Dale's principle, for example, is disregarded. A single cell in the model might sometimes correspond to two cells, or more, in the brain.

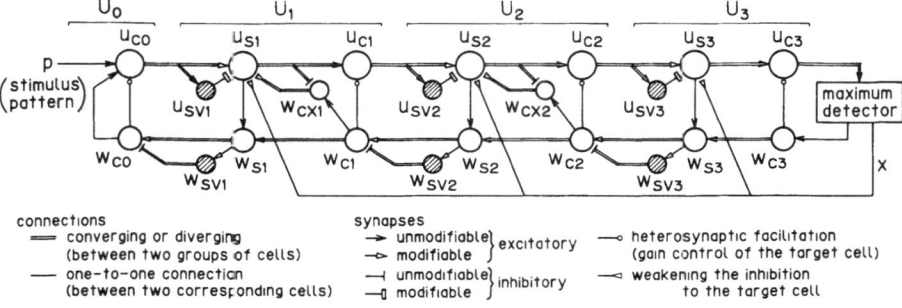

Fig. 9. Hierarchical structure of the interconnections between different kinds of cells in the model for selective attention

transmitted toward higher-order stages through afferent synapses. At the same time, the response of cells of higher stages is fed back to lower-order cell-layers through the efferent synapses. The afferent and the efferent signals interact with each other at every cell-layer: the efferent signals have a facilitating effect on the afferent ones, and, at the same time, the afferent signals gate efferent signal flow.

3.2. Cells in the Efferent Path

Let us now consider a case in which a stimulus is presented to the network. Let one of the gnostic cells (or u_{CL}-cells) in the highest stage of the network be activated by the afferent signals, and one of the patterns in the stimulus recognized. Then, efferent signals, namely, the signals for selective attention, are sent back towards the input layer from that gnostic cell. Efferent signals are transmitted to lower-order cells through w_S- and w_C-cells, which make pairs with u_S- and u_C-cells in the afferent path respectively (Fig. 9). Here, the efferent signals are made to retrace the same paths as the afferent signals in the opposite direction.

Let us first consider the efferent paths from an arbitrary w_S-cell to the w_C-cells of the preceding stage (Fig. 10). During the process of learning, it is assumed that the self-organization of afferent synapses to u_S-cells comes first in a similar manner as in the neocognitron, and that the efferent synapses from w_S-cells are reinforced later. It is further assumed that there exists a certain mechanism to control the strength of the efferent synapses in the following manner: the efferent synapses descending from a w_S-cell finally come to have a strength proportional to the afferent synapses ascending to the u_S-cell which makes a pair with the w_S-cell. Even the efferent synapse to the w_{SV}-cell, a subsidiary inhibitory cell corresponding to the u_{SV}-cell in the afferent paths, is also reinforced in proportion to the afferent inhibitory synapse from the u_{SV}-cell to the u_S-cell. Hence, the efferent signals descending from the w_S-cell are always transmitted in the opposite direction through the same paths as the afferent signals ascending to the u_S-cell. The inhibitory efferent signals via the w_{SV}-cell are also transmitted in the opposite direction through the same paths as the afferent signals ascending via the u_{SV}-cell.

As a result of the network structure described above, if an excitatory afferent path is formed to a u_S-cell from a u_C-cell, an excitatory efferent path comes to be formed automatically from the corresponding w_S-cell to the corresponding w_C-cell. When the afferent in-

hibitory path via the inhibitory u_{SV}-cell is stronger than the direct excitatory path from the u_C-cell, the efferent inhibitory path to the w_C-cell via the inhibitory w_{SV}-cell becomes stronger than the direct excitatory path from the w_S-cell, and the w_C-cell comes to receive an inhibitory overall effect from the w_S-cell. In short, depending on whether a u_S-cell is afferently receiving an overall excitatory or inhibitory effect from a u_C-cell, the corresponding w_C-cell also receives efferently an overall excitatory or inhibitory effect from the corresponding w_S-cell.

We will next consider the efferent signals from w_C-cells to w_S-cells. Corresponding to the afferent synapses which converge to a u_C-cell from a number of u_S-cells, many efferent synapses diverge from a w_C-cell towards w_S-cells (Fig. 10). It is not desirable, however, for all the w_S-cells which receive excitatory efferent signals from w_C-cells to be activated. The reason is as follows: to activate a u_C-cell, the activation of at least one preceding u_S-cell is enough, and usually only a small number of preceding u_S-cells are actually activated. In order to elicit a similar response from the w_S-cells in the efferent paths, the network is synthesized in such a way that each w_S-cell receives not only excitatory efferent signals from w_C-cells but also a gate signal from the corresponding u_S-cell; and it is activated only when it receives a signal both from u_S- and w_C-cells.

Because of this network architecture, in the efferent paths from w_C-cells to w_S-cells, the signals retrace the same route as the afferent signals from u_S-cells to u_C-cells.

3.3. The Action of Efferent on Afferent Signals

As described above, efferent signal flow is affected by afferent signals. Efferent signals, however, are not only affected by afferent signals, but also influence the afferent signal flow.

Let us consider a case in which a stimulus consisting of two patterns or more is presented to the input layer. Let one of the gnostic cells u_{CL} in the highest stage of the network be activated by the afferent signals, and one of the patterns in the stimulus recognized. As described above, efferent signals descending from the activated u_{CL}-cell are now sent only to cells relevant to the recognition of the pattern now recognized. In the intermediate stages of the afferent path, however, not only cells transmitting signals relevant to the pattern now recognized at the highest stage, but also cells transmitting signals for other patterns are usually activated. In order to preserve only the output of cells relevant to the recognition of the finally-recognized pattern and to suppress the output of the other cells, a synaptic connection transmitting a facilitating signal is formed from each w_C-cell to corresponding u_C-cell (Figs. 9 and 10). When the facilitating signal is not sent to the u_C-cell, the gain between the inputs and the output of the cell is gradually attenuated with the passage of time by an effect like "habituation". When a facilitating signal is given to the cell, however, the attenuated gain is forced to recover, and attenuation does not occur. Consequently, u_C-cells in the paths in which efferent signals are flowing do not have their gain attenuated.

Because of this network structure, even if a complex figure consisting of two patterns or more is presented, only the afferent signals relevant to the pattern now recognized are facilitated by the action of efferent signals from the gnostic cell. On the other hand, the afferent signals corresponding to other patterns are gradually attenuated because they receive no facilitation. This means that attention is selectively paid to only one of the patterns in the stimulus.

Incidentally, this effect of facilitation corresponds, for example, to the heterosynaptic facilitation observed in *Aplysia*[2].

When an imperfect or greatly deformed pattern is presented to the input layer, the feature-extracting cells u_S in the afferent path might fail to respond to the features of the pattern. If activated u_S-cells are not found in the afferent path, efferent signals cannot descend any farther, because the efferent signal-flow is guided by the afferent signals. In such a case, the threshold of the u_S-cells is lowered, and the u_S-cells are forced to respond even to incomplete or vague traces of the features. If only these features are thus detected by u_S-cells, the efferent signals now come to be further transmitted to the lower stages via the w_S-cells corresponding to the activated u_S-cells. This mechanism is discussed below in more detail.

Consider a case in which a u_C-cell is silent while the corresponding w_C-cell is activated by efferent signals. This means that the feature which the u_C-cell should extract has not been detected in the afferent paths. In other words, none of the u_S-cells preceding this u_C-cell could extract the feature which is supposed to exist there. It is a w_{CX}-cell that detects the situation in which u_S-cells do not yield an output.

In order to detect a condition in which a w_C-cell, but not a u_C-cell, is activated, one might think that the subtraction of the output of the u_C-cell from the output of the w_C-cell would be enough. However, this is not

always the case, because the output of the u_C-cell might be attenuated by habituation. Hence, in the present model, instead of the output of the u_C-cell itself, the signal converging from the preceding u_S-cells is used (Fig. 10).

Now, if a w_{CX}-cell is activated, it releases a kind of neuromodulator x_S to corresponding u_S-cells. In this case, the neuromodulator is transmitted through paths inverse to the afferent connections converging to the u_C-cell from the u_S-cells.

The neuromodulator x_S weakens the efficiency of the inhibitory signals from the u_{SV}-cell, and decreases the selectivity of feature-extraction by the u_S-cell. Incidentally, a similar process has been hypothesized in the hierarchical associative memory model by Fukushima[6].

Thus, the neuromodulator from w_{CX}-cells makes u_S-cells respond even to incomplete features to which, in the normal state, no u_S-cell would respond. In other words, by control of the neuromodulator, u_S-cells search for even vague traces in defective parts of the stimulus pattern, and try to detect features which should exist there.

Once a u_S-cell has been thus activated, the efferent signal now comes to be further transmitted to the lower stages via the w_S-cell corresponding to the u_S-cell.

3.4. Association and Segmentation (Initial Stage U_0)

As described above, even if a number of stimulus patterns are simultaneously presented to the network, only components of the recognized pattern, determined by the response of u_{CL}-cells in the highest stage, are fed back to layer W_{C0} (*i.e.*, the layer of w_C-cells in the initial stage). In this case, the efferent signals retrace the same route as the afferent ones. Hence, even if the stimulus pattern is a deformed version of the standard pattern which has already been learned, only the signal components corresponding to the recognized pattern appear in layer W_{C0} for the identically deformed shape. If this deformed stimulus pattern has some parts missing, the network tries to detect even the slightest trace of the missing components by automatically lowering the threshold of feature-extracting cells near the defects. Thus defective parts, if not too large, are interpolated, and a complete pattern emerges in layer W_{C0}. In contrast to this, no efferent signals come back for components of noise or stains in the stimulus, and the pattern from which the noise and stains have been eliminated appears in layer W_{C0}.

Therefore, the output of layer W_{C0} can be interpreted as an auto-associative recall from the associative memory. From a different point of view, the output of layer W_{C0} can also be interpreted as the result of segmentation, where only components relevant to a single pattern are selected from the stimulus. The category of the segmented pattern can be judged (or recognized) by noting which u_{CL}-cell is activated in the highest stage. Thus, we can also conclude that segmentation and pattern recognition are simultaneously performed.

In order to improve the ability to interpolate defects, the network has such a structure that the recalled output of layer W_{C0} is fed back positively to layer U_{C0}. That is, the stimulus input p and the recalled output of W_{C0} are superposed in layer U_{C0}, as in Fig. 1.

Since such a positive feedback loop is included in the network, the recalled output, which is first generated only by the stimulus p, is fed back to the input, and another recalling process goes on with the newly-recalled output as a new input. Therefore, even if the interpolation cannot be completed in a single recalling process because the defects are too large, the defects will gradually be filled up during a chain of associative recall, and a perfect pattern without any defect is finally obtained.

Incidentally, not only the feedback signal from W_{C0} but also the stimulus input p is still given to layer U_{C0}, even after the start of the chain of associative recall. This mechanism prevents unlimited drift in the recall, and suppresses the appearance of useless patterns unrelated to the stimulus.

3.5. The Maximum-Detector

When two or more patterns are simultaneously presented, two or more u_C-cells may be activated simultaneously, even in the highest stage U_L. In such a case, the stimulus pattern corresponding to the u_{CL}-cell of maximum output is selected first, so as to start processing the patterns in turn. This selection of the maximum-output cell is done by the maximum-detector shown at far right in Fig. 1. The maximum-detector selects the one which is yielding the largest output from among the u_{CL}-cells. The pattern of the category which corresponds to the u_{CL}-cell thus selected can now be interpreted as recognized.

Once the maximum-output u_{CL}-cell has been selected in the highest stage, only the w_{CL}-cell which makes a pair to this u_{CL}-cell is activated by the excitatory signal from the maximum-detector, and the other w_{CL}-cells stay in a resting state.

Therefore, in the network, only afferent paths relevant to the pattern which is now being detected (or recognized) at the highest stage are facilitated by the action of efferent signals from the activated w_{CL}-cell. The mechanism of this facilitation has already been discussed.

Sometimes all the u_{CL}-cells in the highest stage are silent. For instance, if too many patterns are simultaneously presented, or if they are too close to one another, the signals from the different patterns hinder each other, and are unable to elicit any response from the u_{SL}-cells. Hence u_{CL}-cells which receive signals from u_{SL}-cells are also silent. Even when presenting a single stimulus pattern, there is also a chance that no u_{CL}-cell will be activated if the stimulus pattern is greatly different in shape from the original pattern, or if it has large defects or a lot of noise.

The situation in which all u_{CL}-cells are silent is detected by the maximum-detector, and the maximum-detector sends out signal x through the path shown in Fig. 9. This x-signal is commonly distributed to all the feature-extracting cells u_S of all stages.

Feature-extracting cells u_S receive two kinds of neuromodulators: x_S released by the w_{CX}-cells, which was mentioned before, and x_X released by signal x from the maximum-detector. The neuromodulator x_X, together with x_S, weakens the efficiency of inhibition of the feature-extracting cells u_S, lowers the threshold, and decreases the selectivity of their response. As the result, feature-extracting cells respond easily, even to slightly imperfect features or to stimuli mixed with irrelevant features.

The influence of the signal x on u_S-cells is not identical at all stages, but is greater for u_S-cells in higher stages. This is useful because of the following. Let two or more patterns be simultaneously presented. In the higher stages, each u_S-cell has a greater chance of being presented with irrelevant features together with the feature which the cell is due to extract because it has a larger receptive field. Hence the selectivity of response should be adaptively changed to extract the relevant feature without being hindered by accompanying irrelevant features. In the lower stages, however, the control of selectivity is not so important, because each u_S-cell has a smaller receptive field, and a smaller chance of being presented with two or more features simultaneously.

The longer the silent state of the u_{CL}-cells continues, the more is the accumulation of neuromodulator x_X released by the maximum-detector. Hence, at least one u_{SL}-cell, and consequently one u_{CL}-cell, is activated after a certain time. Once a u_{CL}-cell is activated, the release of neuromodulator x_X stops, and the neuro-modulator begins to decay.

3.6. Switching Attention

Suppose one of the multiply-presented stimulus patterns is being attended to and recognized. In order to switch attention to another pattern, a momentary interruption of the efferent signal-flow is enough. Each u_C-cell, after the disappearance of the facilitating signal from the corresponding w_C-cell in the efferent path, will have its gain lowered if this has so far been kept high by the facilitating signal. A decrease in gain occurs like fatigue, depending on the degree of the forced increase of gain until then. On the other hand, the u_C-cell will recover its gain, if this has so far been attenuated by "habituation".

Since the gain of the u_C-cells is thus controlled after the interruption of the efferent signal flow, signals corresponding to other patterns in the stimulus can flow more easily through the afferent paths, and usually another gnostic cell, which has hitherto been silent, will be activated. Consequently, the afferent paths for the new pattern will be facilitated by the efferent signals from the newly-activated gnostic cell.

By means of a repetition of this process, attention is switched to each of the patterns in the stimulus figure in turn, and they are recognized and recalled one after another.

3.7. Computer Simulation

We will now study the behaviour of our network by computer simulation. The scale of the simulated network is $L = 3$.

We will discuss mainly the response of a network which has already finished learning, but we shall also briefly describe the earlier process of learning. In the initial state before learning, all the modifiable synapses of the network were assumed to have a strength of zero. During the learning period, five training patterns, "0", "1", "2", "3" and "4" shown in Fig. 11, were repeatedly presented as a stimulus input p, and the self-organization of the network occurred in a similar manner as in the neocognitron. In the simulation, it was assumed that the self-organization of afferent synapses came first when all efferent signal-flow was stopped,

and that the efferent synapses were reinforced later. Habituation of the u_C-cells was assumed not to occur during the self-organization. It was further assumed that all the modifiable synapses lost their "plasticity" after finishing learning.

Figs. 12–14 shows the behaviour of a network which has already finished learning. In these figures, the response of the cells in layers U_{C0} and W_{C0} of the initial stage is shown in time sequence. The intensity of response of the cells, which are arranged in a two-dimensional array, is shown by the density of the photographs. The patterns in the upper line show the response of layer U_{C0}, and the patterns in the lower line show the response of layer W_{C0}. The numeral to the upper left of each pattern represents time t after the start of stimulus presentation. The mark ▼ to the right of the numeral denotes that the efferent signal-flow has been momentarily interrupted just before this moment to switch attention. The stimulus pattern given to this network is identical to the response of layer U_{C0} at $t = 0$, which is shown in the upper left of each figure. (It should be noted that the input pattern p appears directly in layer U_{C0} at $t = 0$, because no response is elicited from layer W_{C0} at $t < 0$).

Fig. 12 is the response to the stimulus "20" shown in the upper left. In layer U_{CL} at the highest stage of the network, which is not shown in this figure, the gnostic cells corresponding to pattern "2" and "0" are both activated at $t = 0$, but a larger response is elicited from the cell corresponding to "0". Hence, in layer W_{CL}, only the cell corresponding to "0" is activated by the signal from the maximum-detector. This signal is fed back further to layer W_{C0} through the efferent paths. With the lapse of time, habituation of the u_C-cells builds up gradually in each layer. At $t = 3$, the components of pattern "2", which is not being attended to, have already become quite weak.

At this time, in order to switch attention, the efferent signal-flow is interrupted for a moment, and the habituation of the cells recovers. The new response thus elicited is shown at $t = 4$. In layer U_{CL}, only the gnostic cell for pattern "2" is now activated, because the gain of the u_C-cells which have so far been responding strongly to pattern "0" is now attenuated by fatigue. Hence, also in layer W_{C0} in the initial stage, only the pattern components of "2" appear.

Although patterns "0" and "2" in the stimulus are a little different in shape from the training patterns (Fig. 11), the patterns fed back to layer W_{C0} are identical in shape with the stimulus patterns now presented.

Fig. 11. Five training patterns used for the self-organization of the model for selective attention

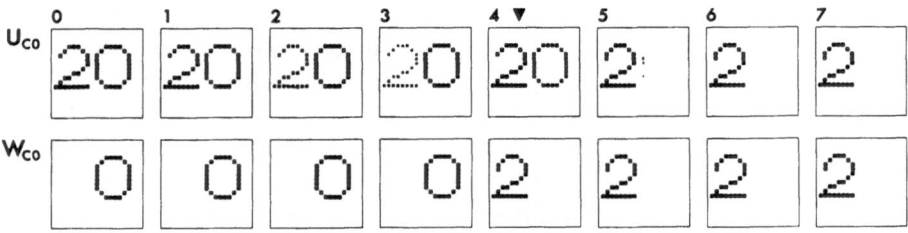

Fig. 12. An example of the response of the model to juxtaposed patterns

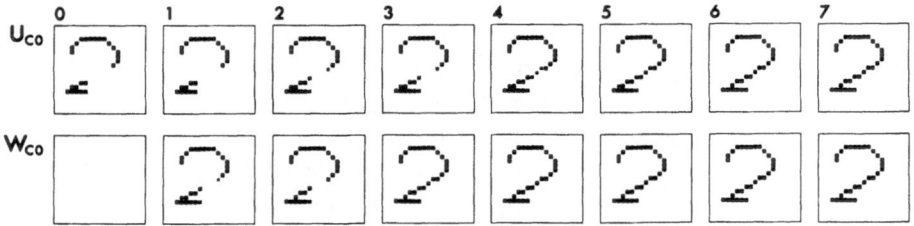

Fig. 13. An example of the response of the model to an incomplete distorted patterns

Fig. 13 shows the response to a greatly deformed pattern which even has one part missing. Since the difference between the stimulus and the training pattern is too large, no response is elicited from layer u_{CL} (not shown in Fig. 13), and accordingly, no feedback signal appears at layer W_{C0} at $t = 0$. This situation is detected by the maximum-detector, and signal x is sent. At time $t = 1$, the gnostic cell for "2" becomes activated in the highest stage, because the threshold of the u_S-cells in the network is decreased by the x-signal. In the pattern which is now fed back to layer W_{C0}, both ends of the missing part are already beginning to be interpolated. Partly interpolated, this signal, namely the output of layer W_{C0}, is given to layer U_{C0} again, together with the stimulus p. The interpolation continues gradually while the signal is circulating in the feedback loop, and finally the missing part of the stimulus is completely filled in. As may be seen in Fig. 13, the missing part is interpolated quite naturally, even though the difference in shape between the stimulus and the training pattern is considerable. Incidentally, in the pattern for which interpolation is already finished, the horizontal bar at the bottom of the "2" is shorter than in the training pattern. But however short the horizontal bar is, the pattern is still a perfect character "2". Hence, this component of the pattern is left intact and is reproduced like the stimulus pattern. Thus, the deformation of the stimulus pattern is tolerated as it is, and only indispensable missing parts are naturally interpolated without any strain.

Fig. 14 shows an example of the reponse to a stimulus consisting of superposed patterns. As may be seen in the figure, the pattern "4" is isolated first, the pattern "2" next, and finally pattern "1" is extracted.

4. Discussion

The hierarchical structure of the neocognitron was originally suggested by the classical hypothesis of Hubel and Wiesel[8, 9], according to which the neural network in the visual (or striate) cortex has a hierarchical structure: LGB (lateral geniculate body) → simple cells → complex cells → lower-order hypercomplex cells → higher-order hypercomplex cells. They also hypothesized that the neural network between lower-order hypercomplex cells and higher-order hypercomplex cells has a structure similar to the network between simple cells and complex cells. In other words, networks of similar structure are connected in a hierarchical cascade. In this hierarchy, simple cells and lower-order hypercomplex cells correspond to the u_S-cells in the neocognitron, and complex cells and higher-order hypercomplex cells to the u_C-cells. It is true, however, that the hierarchical model of Hubel and Wiesel must be modified a littel from its original form. In fact, several experiments contradict their hierarchical model, such as the discovery of monosynaptic connections from LGB to complex cells. Even the existence of hypercomplex cells is doubted. Thus, the original hypothesis of Hubel and Wiesel is questionable if we restrict our discussion to the neural network within the striate area only. However, if we consider the information flow in the whole visual system of the brain, there is certainly a hierarchical structure between

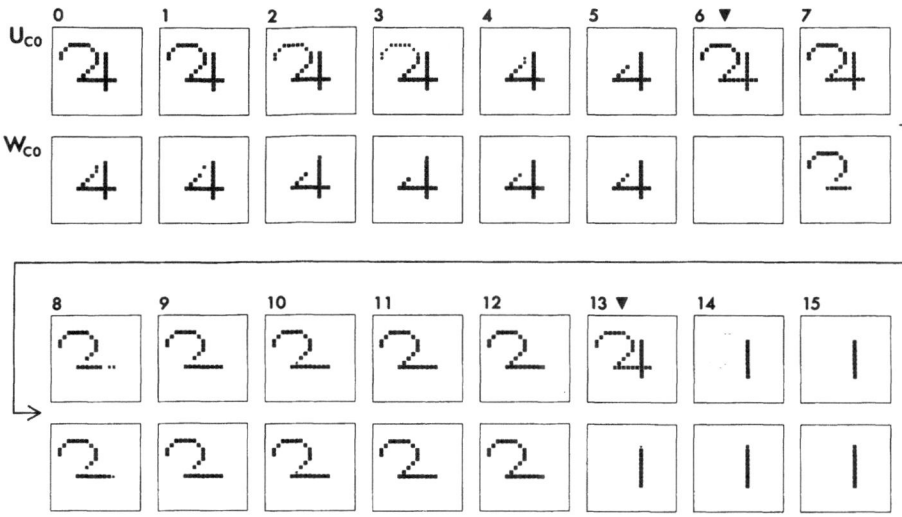

Fig. 14. An example of the response of the model to superposed patterns

different cortical areas. Such structure suggests the correctness of the neocognitron model, if we restrict our discussion, not to a one-to-one correspondence between the cells of the model and the cells of the brain, but to a global information flow in the whole visual system.

As we discussed in the text, another hypothesis employed in the neocognitron is the assumption that, even in the process of self-organization, all the cells of each cell-plane come to possess input synapses of the same spatial distribution, in which only the position of the presynaptic cells shift in parallel in accordance with a shift in position of individual postsynaptic cells. It is not known how the modifiable synapses of the nervous system are actually modified under such conditions. However the correctness of the hypothesis is suggested, for example, by the fact that orderly synaptic connections are formed between the retina and the optic tectum not only in the initial development of the embryo, but also in the regeneration in an adult amphibian or fish: for example, in the regeneration which takes place after the removal of half the tectum, the whole retina comes to make an orderly compressed projection on the remaining half tectum (*e.g.* review article by Meyer and Sperry[12]). This suggests that, in the nervous system, there is some mechanism which insures the formation of orderly synaptic connections between groups of cells.

In the second model, it is assumed that the efferent connections descending from a w_S-cell come to have a strength proportional to the afferent connections ascending to the u_S-cell which make a pair with the w_S-cell. In other words, mirror-image point-to-point con-

nections are formed between the afferent and the efferent systems of the network. The correctness of this assumption is suggested by anatomical evidence: for example, Tigges et al.[16] found a mirror-image point-to-point connection between area 18 and area 17. Wong-Riley[17] reports that the major visual cortical areas are interconnected in a precise topographical and reciprocal fashion. There are many other papers reporting the existence of reciprocal connections between other cortical areas as well.

We will next consider the characteristics of the receptive field of cells of the second model. The cells in the afferent paths, especially in the higher stages, have a large receptive field. If two stimuli are presented to the receptive field of a cell, only a stimulus which is being attended to elicits a large response; and the response to a stimulus which is not being attended to is greatly reduced. This coincides closely, for example, with a neurophysiological experiment of Moran and Desimon[13]. They recorded the output of single cells in the prestriate area V 4 and the inferior temporal cortex of monkeys, and found that the response to a stimulus which was not being attended to was considerably reduced.

Appendix: Mathematical Description of the Models

In the following equations, notation u_{Sl}^t (**n**, k), for example, is used to denote the output of a u_S-cell in the l-th stage, where **n** is a two-dimensional set of coordinates indicating the position of the cell's receptive-field centre in the input layer (layer U_0 of the neocognitron or layer U_{C0} of the second model), and k

$(= 1, 2, \ldots, K_l)$ is a serial number indicating the type of feature which the cell extracts. In other words, k is a serial number of the cell-plane. Time t is assumed to take a discrete integer value.

Incidentally, since we have $K_0 = 1$ in stage U_0, the output of a u_C-cell, for example, might be denoted by $u_{C0}^t(\mathbf{n})$, in which k is abbreviated. In the highest stage, since only one u_C- and w_C-cell (in the second model) exist for each value of k, expressions like $u_{CL}^t(k)$, in which $\mathbf{n}$ is abbreviated, are sometimes used to denote the output of such cells.

A. Neocognitron[3, 5]

A.1. Response of the Cells

The output of a feature-extracting cell u_S of stage U_l is given by

$$u_{Sl}^t(\mathbf{n}, k) =$$

$$r_l \cdot \varphi \left[\frac{\sigma_l + \sum_{\kappa=1}^{K_{l-1}} \sum_{v \in A_l} a_l(\mathbf{v}, \kappa, k) \cdot u_{Cl-1}^t(\mathbf{n}+\mathbf{v}, \kappa)}{\sigma_l + \frac{r_l}{1+r_l} \cdot b_l(k) \cdot u_{SVl}^t(\mathbf{n})} - 1 \right]$$

(1).

where

$$\varphi[x] = \begin{cases} x & \text{if} \quad x \geqq 0 \\ 0 & \text{if} \quad x < 0 \end{cases} . \qquad (2)$$

In the case of $l = 1$ in (1), $u_{Cl-1}^t(\mathbf{n}, \kappa)$ stands for the output of a u_0-cell of the input layer $u_0^t(n)$, and we have $K_{l-1} = 1$. Parameter σ_l is a positive constant determining the level at which saturation starts in the input-to-output characteristic of the cell.

$a_l(\mathbf{v}, \kappa, k)$ $(\geqslant 0)$ is the strength of the modifiable excitatory synapse coming afferently from cell $u_{Cl-1}(\mathbf{n}+\mathbf{v}, \kappa)$ in the preceding stage U_{l-1}, and A_l denotes the summation range of $\mathbf{v}$, that is, the size of the spatial spread of the afferent synapses to one u_S-cell. $b_l(k)$ $(\geqslant 0)$ is the strength of the modifiable inhibitory synapse coming from inhibitory cell $u_{SVl}(\mathbf{n})$. As discussed in the text, it is assumed that all the u_S-cells in a cell-plane have identical set of input synapses. Hence, a_l $(\mathbf{v}, \kappa, k)$ and $b_l(k)$ do not contain argument $\mathbf{n}$ representing the position of the receptive field of the cell $u_{Sl}(\mathbf{n}, k)$.

As can be seen from (1), the inhibitory input to this cell acts in a shunting manner. The positive constant r_l determines the efficiency of the inhibitory input to this cell.

The inhibitory cell u_{SV} which sends an inhibitory signal to this u_S-cell yields an output equal to the weighted root-mean-square of the signals from the preceding u_C-cells, that is,

$$u_{SVl}^t(\mathbf{n}) = \sqrt{\sum_{\kappa=1}^{K_{l-1}} \sum_{v \in A_l} c_l(\mathbf{v}) \cdot \{u_{Cl-1}^t(\mathbf{n}+\mathbf{v}, \kappa)\}^2} , \qquad (3)$$

where $c_l(\mathbf{v})$ represents the strength of the excitatory unmodifiable synapses, and is a monotonically decreasing function of $|\mathbf{v}|$, which satisfies

$$\sum_{\kappa=1}^{K_{l-1}} \sum_{v \in A_l} c_l(\mathbf{v}) = 1 . \qquad (4)$$

Incidentally, the role of the root-mean-square cells in feature extraction is discussed elsewhere[4, 5].

The u_C-cells are inserted in the network to allow for positional errors in the features of the stimulus. The output of a u_C-cell is given by

$$u_{Cl}^t(\mathbf{n}, k) = \psi \left[\sum_{v \in D_l} d_l(\mathbf{v}) \cdot u_{Sl}^t(\mathbf{n}+\mathbf{v}, k) \right] , \qquad (5)$$

where $\psi[\ ]$ is a function specifying the characteristic of saturation, and is defined by

$$\psi[x] = \frac{\varphi[x]}{1+\varphi[x]} . \qquad (6)$$

Parameter $d_l(\mathbf{v})$ denotes the strength of the excitatory unmodifiable synapses, and is a monotonically decreasing function of $|\mathbf{v}|$. Hence, if at least one preceding u_S-cell is activated in the area D_l, to which these synapses spread, this u_C-cell is also activated.

A.2. Modifiable Synapses

During the self-organization, the efficiencies of the modifiable synapses $a_l(\mathbf{v}, \kappa, k)$ and $b_l(k)$ are reinforced depending on the strength of the input to the "seed cells". Let $u_{Sl}(\hat{\mathbf{n}}, \hat{k})$ be selected as a seed cell at time t. The modifiable synapses $a_l(\mathbf{v}, \kappa, \hat{k})$ and $b_l(\hat{k})$ to this seed cell are reinforced by the following amount:

$$\Delta a_l(\mathbf{v}, \kappa, \hat{k}) = q_l \cdot c_l(\mathbf{v}) \cdot u_{Cl-1}^t(\hat{\mathbf{n}}+\mathbf{v}, \kappa) , \qquad (7)$$

$$\Delta b_l(\hat{k}) = q_l \cdot u_{SVl}(\hat{\mathbf{n}}) , \qquad (8)$$

where q_l is a positive constant determining the speed of reinforcement.

B. Model for Selective Attention[7]

B.1. Cells in the Afferent Path

Similarly, as in the neocognitron, the output of a feature-extracting cell u_S of stage U_l is given by

$$u_{Sl}^t(\mathbf{n}, k) = r_l^t(\mathbf{n}, k) \cdot$$

$$\varphi\left[\frac{\sigma_l + \sum\limits_{\kappa=1}^{K_{l-1}} \sum\limits_{\nu \in A_l} \varepsilon(\nu, \kappa, k) \cdot u_{Cl-1}^t(\mathbf{n} + \nu, \kappa)}{\sigma_l + \dfrac{r_l^t(\mathbf{n}, k)}{1 + r_l^t(\mathbf{n}, k)} \cdot b_l(k) \cdot u_{SVl}^t(\mathbf{n})} - 1 \right]$$

(9)

This equation is almost the same as that for the neocognitron, except that the efficiency $r_l^t(\mathbf{n}, k)$ of the inhibitory input to each cell is controlled individually. The response of inhibitory cell u_{SV} which sends an inhibitory signal to this u_S-cell is give by the same equation (3) as for the neocognitron.

Modifiable synapses $a_l(\nu, \kappa, k)$ and $b_l(k)$ are reinforced by means of a similar algorithm to that used for learning-without-a-teacher in the neocognitron, when all efferent signal-flow is stopped and habituation of u_C-cells is made not to occur. Thus, each u_S-cell comes to respond selectively to a particular feature of the stimulus. In this paper, however, we shall consider only the state after learning, and assume that the synapses have already lost their modifiability.

The output of a u_C-cell, which is inserted in the network to allow for positional errors in the features of the stimulus, is given by

$$u_{Cl}^t(\mathbf{n}, k) = g_l^t(\mathbf{n}, k) \cdot \psi\left[\sum_{\nu \in D_l} d_l(\nu) \cdot u_{Sl}^t(\mathbf{n} + \nu, k) \right] .$$

(10)

The variable $g_l^t(\mathbf{n}, k)$ denotes the gain of the u_C-cell, and its value is controlled by the signal from the w_C-cell in the efferent path as in (14).

B.2. Cells in the Efferent Path

The output of a w_C-cell in the efferent path is given by

$$w_{Cl}^t(\mathbf{n}, k) = \psi\left[a_l \cdot \left\{ \sum_{\kappa=1}^{K_l+1} \sum_{\nu \in A_{l+1}} a_{l+1}(\nu, k, \kappa) \cdot \right. \right.$$

$$\left. \left. w_{Sl+1}^t(\mathbf{n}-\nu, \kappa) - \sum_{\nu \in A_{l+1}} c_{l+1}(\nu) \cdot w_{SVl+1}^t(\mathbf{n}-\nu) \right\} \right] .$$

(11)

As you may see by comparing this equation to (9), the efferent synapses diverging from a w_S-cell have a strength porportional to the afferent synapses converging to the u_S-cell which makes a pair with the w_S-cell. The ratio of the strength of these synapses is α_l. It is assumed that the strength of the efferent synapses is automatically modified in this way after finishing the reinforcement of the afferent synapses.

The output of the subsidiary inhibitory cell w_{SV} is given by

$$w_{SVl+1}^t(\mathbf{n}) = \frac{r_{l+1}^0}{1 + r_{l+1}^0} \sum_{\kappa=1}^{K_{l+1}} b_{l+1}(\kappa) \cdot w_{Sl+1}^t(\mathbf{n}, \kappa) .$$

(12)

That is, the efferent synapse from w_{Sl+1}- to w_{SVl+1}-cell is the same as the afferent synapse from u_{SVl+1}- to u_{Sl+1}-cell, where only the direction of the signal flow is reversed. The only difference between the efferent and the afferent paths is that the w_{SV}-cell in the efferent paths has a linear input-to-output characteristic, while the u_{SV}-cell in the afferent paths has a root-mean-square characteristic. The parameter r_l^0 in (12) is the initial value of the variable $r_l^t(\mathbf{n}, k)$ in (9), and will be discussed in connection with (17).

The output of a w_S-cell is given by

$$w_{Sl}^t(\mathbf{n}, k) = \min\left[u_{Sl}^t(\mathbf{n}, k), \sum_{\nu \in D_l} d_l(\nu) \cdot w_{Cl}^t(\mathbf{n}-\nu, k) \right] , \quad (13)$$

where $\min[\ ]$ is a function which takes the smaller value of the two arguments.

B.3. The Action of Efferent on Afferent Signals

The gain of a u_C-cell in the afferent paths (see (10)) is given by

$$g_l^t(\mathbf{n}, k) = \gamma_l \cdot g_l^{t-1}(\mathbf{n}, k) + (1 - \gamma_l) \cdot \psi\left[\gamma_l' \cdot w_{Cl}^{t-1}(\mathbf{n}, k) \right] , \quad (14)$$

where γ_l is a constant determining the speed the habituation of the u_C-cells, and lies in the range $0 < \gamma_l \leqslant 1$. The positive constant γ_l' determines the signal level at which the saturation of facilitation begins. Incidentally, the gain at the initial state of the network is 1.

The output of a w_{CX}-cell, which detects the situation in which u_S-cells fail to extract a feature, is given by

$$w_{CXl}^t(\mathbf{n}, k) = \varphi\left[w_{Cl}^t(\mathbf{n}, k) - \sum_{\nu \in D_l'} d_l'(\nu) \cdot u_{Sl}^t(\mathbf{n} + \nu, k) \right] .$$

(15)

The spreading area D'_l of the afferent synapses $d'_l(\mathbf{v})$ to a w_{CX}-cell from u_S-cells is a little wider than the spreading area D_l of the afferent synapses $d_l(\mathbf{v})$ to a u_C-cell. This is for the following reason: let a stimulus pattern be a slightly deformed version of a learned pattern. The output of the u_C-cells elicited by the stimulus pattern itself might possibly arise at a slightly different position than that of the w_C-cells, which is determined by the response to the training pattern during the learning period. In order to prevent a spurious output from the w_{CX}-cells caused by a discrepancy between the output of these two groups of cells, the spreading area D'_l is a little larger than D_l in (10).

The amount of neuromodulator x_S which acts on cell $u_{Sl}(\mathbf{n}, k)$ is

$$x^t_{Sl}(\mathbf{n}, k) = \beta_l \cdot x^{t-1}_{Sl}(\mathbf{n}, k) + \beta'_l \cdot \sum_{\mathbf{v} \in D_l} d_l(\mathbf{v}) \cdot w^{t-1}_{CXl}(\mathbf{n}\text{-}\mathbf{v}, k) \quad . \tag{16}$$

In other words, neuromodulator x_S is accumulated by an amount proportional to the output of the w_{CX}-cells, but, at the same time, decreases with an attenuation constant β_l ($0 < \beta_l \leqslant 1$). In the computer simulation, we put $\beta_l = 1$. β'_l is another positive constant.

The efficiency of inhibition is determined by the variable $r^t_l(\mathbf{n}, k)$ in (9), and its value is controlled by the neuromodulator x_S as follows:

$$r^t_l(\mathbf{n}, k) = \frac{r^0_l}{1 + x^t_{Sl}(\mathbf{n}, k) + x^t_{Xl}} \quad , \tag{17}$$

where r^0_l is the initial value of $r^t_l(\mathbf{n}, k)$. The variable x^t_{Xl} represents the amount of another neuromodulator released by signal x from the maximum-detector, which is given by (20). In the highest stage U_L, no w_{CX}-cells exist, but equation (17) can be applied if x_S is assumed to be zero.

B. 4. Association and Segmentation (Initial Stage U_0)

The output of a u_{C0}-cell in the initial stage is different from that of other stages, and is given by:

$$u^t_{C0}(\mathbf{n}) = g^t_0(\mathbf{n}) \cdot \max[p^t(\mathbf{n}), w^t_{C0}(\mathbf{n})] \quad . \tag{18}$$

The gain $g^t_0(\mathbf{n})$ is given by (14) and (21) in the same manner as for the intermediate stages. Incidentally, stage U_0 contains only u_C- and w_C-cells, and the output of a w_C-cell is also given by (11).

B. 5. The Maximum-Detector

The output of a w_C-cell of stage U_L is different from that of other stages, and is given by

$$w^t_{CL}(k) = \begin{cases} 1 & \text{if } u^t_{CL}(k) = \max_\kappa [u^t_{CL}(\kappa)] > 0 \\ 0 & \text{else} \end{cases} \tag{19}$$

When the maximum-detector detects the situation in which all the u_{CL}-cells are silent, it sends out signal x. The amount of neuromodulator x_X released by this x-signal to a single u_S-cell is

$$x^t_{Xl} = \begin{cases} x^t_{Xl} + \beta_{Xl} & \text{if } u^t_{CL}(\kappa) = 0 \text{ for all } \kappa \\ \beta'_{Xl} \cdot x^{t-1}_{Xl} & \text{else} \end{cases} \tag{20}$$

In other words, if all the u_{CL}-cells in the highest stage are silent, x_X accumulates by a constant amount β_{Xl}. If at least one u_{CL}-cell is activated, however, it decreases with an attenuation ratio β'_{Xl}.

Incidentally, during the period in which the x-signal is being sent from the maximum-detector, the gain $g^t_l(\mathbf{n}, k)$ of the u_C-cells in the network is assumed to be unchanged and not attenuated by (14).

B. 6. Switching Attention

The gain just after an interruption in the efferent signal-flow is given, not by (14), but by

$$g^t_l(\mathbf{n}, k) = \frac{1}{1 + \gamma''_l \cdot g^{t-1}_l(\mathbf{n}, k)} \quad , \tag{21}$$

where γ''_l is a positive constant.

Incidentally, the amount of neuromodulator x_S and x_X is also assumed to be reset at zero just after the interruption of the efferent signal flow.

References

1. Bruce C, Desimone R, Gross CG (1981) Visual properties of neurons in a polysensory area in superior temporal sulcus of the macaque. J Neurophys 46: 369–384
2. Castelluci V, Kandel ER (1976) Presynaptic facilitation as a mechanism for behavioral sensitization in *Aplysia*. Science 194: 1176–1178
3. Fukushima K (1980) Neocognitron: a self-organizing neural network model for a mechanism of pattern recognition unaffected by shift in position. Biol Cybern 36: 193–202
4. Fukushima K (1981) Cognitron: a self-organizing multilayered neural network model. NHK Technical Monograph No. 30. Tokyo: NHK Tech Res Labs

5. Fukushima K, Miyake S (1982) Neocognitron: a new algorithm for pattern recognition tolerant of deformations and shifts in position. Pattern Recognition 15: 455–469

6. Fukushima K (1984) A hierarchical neural network model for associative memory. Biol Cybern 50: 105–113

7. Fukushima K (1986) A neural network model for selective attention in visual pattern recognition. Biol Cybern 55: 5–15

8. Hubel DH, Wiesel TN (1962) Receptive fields, binocular interaction and functional architecture in cat's visual cortex. J Physiol (London) 160: 106–154

9. Hubel DH, Wiesel TN (1965) Receptive fields and functional architecture in two nonstriate visual area (18 and 19) of the cat. J Neurophysiol 28: 229–289

10. Hubel DH, Wiesel TN (1977) Functional architecture of macaque monkey visual cortex. Proc Roy Soc London, Ser B 198: 1–59

11. Kohonen T (1972) Correlation matrix memories. IEEE Trans Comput C-21: 353–359

12. Meyer RL, Sperry RW (1974) Explanatory models for neuroplasticity in retinotectal connections. In: Stein DG, Rosen JJ, Butters N (eds) Plasticity and function in the central nervous system. Academic Press, New York San Francisco London, pp 45–63

13. Moran J, Desimone R (1985) Selective attention gates visual processing in the extrastriate cortex. Science 229: 782–784

14. Nakano K (1972) Associatron—a model of associative memory. IEEE Trans Syst Man Cybern SMC-2: 380–388

15. Sato T, Kawamura T, Iwai E (1980) Responsiveness of inferotemporal single units to visual pattern stimuli in monkeys performing discrimination. Exp Brain Res 38: 313–319

16. Tigges J, Spatz WB, Tigges M (1973) Reciprocal point-to-point connections between parastriate and striate cortex in the squirrel monkey (*Saimiri*). J Comp Neurol 148: 481–490

17. Wong-Riley M (1979) Columnar cortico-cortical interconnections within the visual system of the squirrel and macaque monkeys. Brain Res 162: 201–217

Correspondence: Kunihiko Fukushima, Ph.D., NHK Science and Technical Research Laboratories, 1-10-11, Kinuta, Setagayaku, Tokyo 157, Japan.

Acta Neurochirurgica, Suppl. 41, 68–77 (1987)

Visual Cortical Plasticity in Infant Kittens

K. Toyama and **Y. Komatsu**

Department of Physiology, Kyoto Prefectural School of Medicine, Kyoto, Japan

Summary

The visual cortex of the cat is characterized by marked modifiability of neuronal responsiveness by visual experience in infancy, and stereotyped pattern of functional architectures in adulthood. The question of how the plasticity of the infant visual cortex is compatible with the regular patterns of the adult visual cortex has been a central problem of the brain neuroscience. This question was answered by quantifying the plasticity in the visual cortical circuitry of the infant kittens as changes in synaptic transmission produced after conditioning stimulation of the visual pathway. The results indicate that the solution of this question is the heterogenous distribution of the synaptic plasticity in the infant visual cortex: the plasticity is not uniformly present in the visual cortical circuitry, but is limited only to a part of the circuitry (the cortico-cortical synapses in the supragranular layers). Therefore, the visual function (photic responsiveness) may be learned during the postnatal life of the kittens by the supragranular cells with plastic synapses, while the other cortical cells with fewer plastic synapses put prenatally designed constraints on the learning, so that the learning yields the adult cortical circuitry with regular patterns of organization.

Keywords: Visual cortical plasticity; visual cortical development; visual cortex; synaptic plasticity; visual learning; long-lasting potentiation.

Introduction

The dispute between Nature versus Nurture theory concerning visual cortical plasticity has been one of the most exciting topics in neuroscience. The present study provides evidence that synaptic plasticity is not uniformly distributed but rather restricted to a part of the visual cortical circuitry, which may provide an important clue to answer the Nature-Nurture question.

In the dispute during the last two decades it has been established that photic responsiveness of visual cortical cells is still immature in newborn kittens and develops under the influence of the visual environment (Hubel and Wiesel, 1963; Blakemore and van Sluyter, 1975; Buisseret and Imbert, 1976). For example, the number of cells with orientation selectivity—the most important response properties of the adult cells providing the basis for visual perception—is very small at birth, less than 20%. It increases rapidly between 2 and 4 weeks and reaches the adult level several weeks after birth (Fig. 1 A). If visual experience is restricted to a certain visual pattern during the period of rapid development in orientation selectivity, the selectivity develops only to that pattern (Hirsch and Spinelli, 1970; Blakemore and Cooper, 1970; Blakemore et al., 1973; Blasdel, 1977; Rauschecker and Singer, 1981; Singer et al., 1981). It was shown that in a kitten raised in a cylinder with vertical stripes for two months after birth (Fig. 1 B), most visual cortical cells exhibit vertical orientation selectivity (bars in Fig. 1 D), while those in a kitten exposed to horizontal stripes mostly exhibited horizontal orientation selectivity (Fig. 1 C) (Cooper and Blakemore, 1970). Therefore, it is as if the cortical cells learned orientation selectivity by exposure to the visual environment. This is the case only for a critical period extending between 3–8 weeks: they learn only during that period and not before or after this period (Hubel and Wiesel, 1973).

In spite of general agreement on phenomenology in development and modification of neuronal responsiveness in the visual cortex (VC), its neuronal basis is still a subject of controversy. Nurture theory assumes that selectivity is learned from visual experience (Hebb, 1949; Blakemore and Cooper, 1970; Blakemore and Mitchell, 1973; Blakemore and van Sluyter, 1975; Rauschecker and Singer, 1981; Singer et al., 1981), while Nature theory assumes that selectivity is innately determined, but undergoes selection by visual experience, that is, there is already a prototype of

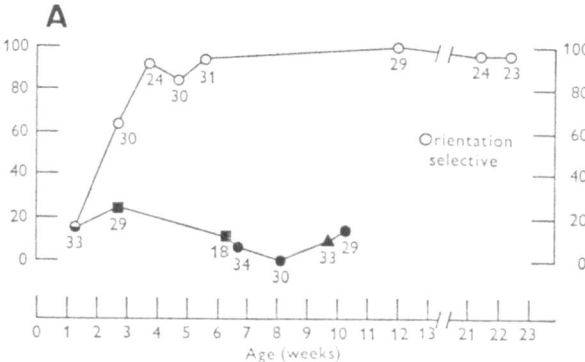

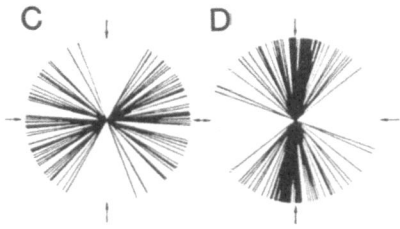

Fig. 1. Plasticity in neuronal responsiveness in the visual cortex of infant kittens. A) represents development of orientation selectivity in normal (open circles) and in dark-reared condition (filled triangles and squares) (modified from Blakemore and van Sluyters, 1975). B)–D) represents effects of visual environment on orientation selectivity. B) schematic illustration of the cage to expose vertical bars to a kitten. C) orientation preference of cells recorded from the visual cortex of kittens that experienced horizontal bars. D) similar to C) but for kitten that experienced vertical bars (Blakemore and Cooper, 1970)

selectivity in newborn kittens and it develops into adult selectivity only in cells that experienced visual stimuli coincident with the prototype selectivity (Hubel and Wiesel, 1963; Wiesel and Hubel, 1963, 1965; Stryker et al., 1978). The simplest neural model for the former hypothesis may be that visual cortical synapses are newly formed under the influence of visual experience

according to the Hebbian algorithm (Hebb, 1949), while the model for the latter hypothesis may be that pre-existing but unfunctional synapses are activated by visual experience.

It has been shown in the adult cat (Toyama et al., 1974) that visual signals are transferred from the lateral geniculate nucleus (LGN) to layer IV in VC through geniculocortical synapses, and they are further transferred from there to layer II through cortico-cortical synapses (Fig. 2 B). Probably the signals are transferred further on to layers III and V, which are the main stations for cortico-cortical and subcortical projection (Fig. 2 A). Knowledge of synaptic plasticity at every stage of synaptic transmission in the main cortical circuitry is essential for determining which hypothesis is true.

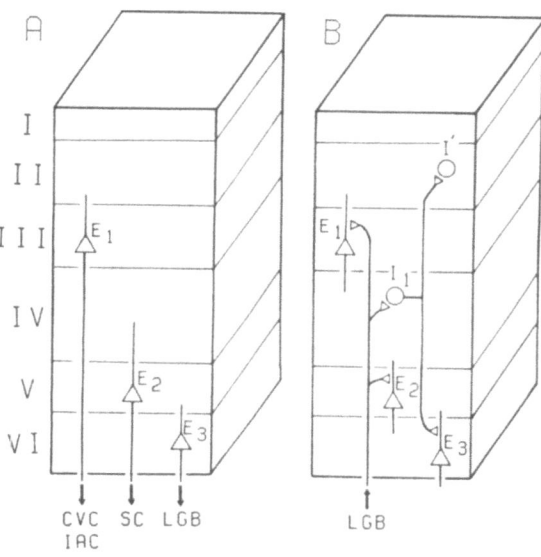

Fig. 2. Visual cortical circuitry in adult cats. A) corticofugal pathway. B) excitatory connection. E_{1-3}, efferent cells, I_I and I', excitatory interneurons (Toyama et al., 1974)

Our strategy for doing this was to simplify the preparation. Studies of visual cortical plasticity have been made so far by assessing the plastic changes rather indirectly through modification of neuronal responsiveness by visual experience. This has certainly been quite helpful in demonstrating the incredible plasticity in visual cortical circuitry. However, assessing plasticity through response properties has some disadvantages. First, it provides no qualitative measure of synaptic plasticity involved in visual cortical circuitry. It indicates that the circuitry has been modified, but not how much or where. Second, there is a difficulty in controlling the visual inputs to evoke the plastic changes. For example, in the case of Cooper and

Blakemore, it may be argued that vertical stripes may be seen as horizontal if the animal's head is tilted 90°. There is compensatory eye rotation for the body tilt, but this was found to be insufficient, and now a better way to control visual experience is to present stripes with goggles. Even this method has difficulty in controlling the alertness of the animal which may be important in visual learning.

It is therefore desirable to design an experiment where all experimental parameters are under control. This was accomplished in our work (Komatsu et al., 1981) by using a slice preparation of VC, and by replacing the visual experience with electrical stimulation of the white matter (WM). The plastic changes were assessed as potentiation of synaptic transmission that occurred after conditioning stimulation. Three indices were used to assess synaptic transmission: amplitude of field potentials (FPs), orthodromic latency of extracellularly recorded responses of cortical cells, and local currents demonstrated by current source-density (CD) analysis of the FPs.

Methods

Coronal sections (0.4–0.5 mm in thickness) were dissected from the VC (area 17) of infant kittens. The cortical slice (bs in Fig. 3 A) was placed in a recording chamber perfused with Krebs-Ringer solution saturated with a mixture of 95% O_2 and 5% CO_2. A pair of stimulating electrodes (S_1 and S_2 in Fig. 3 A) were placed in WM. A train of pulses at 2 Hz was applied for one hour as a conditioning stimulus through S_1, while single pulses were applied through S_2 at 0.2 Hz as test stimuli to assess the changes in synaptic transmission after conditioning stimulation. A glass microelectrode (m_1) was inserted in the cortical slice to record extracellularly FPs and unit responses evoked by test stimulation, and another microelectrode (m_2) was placed in WM to monitor the effects of conditioning and test stimulation.

Results

1. Analysis of Orthodromic Latencies in Slice Preparations

Fig. 4 A illustrates FPs in VC and presynaptic volleys in WM evoked by test stimulation before, and two hours after conditioning stimulation. The FPs were almost doubled after conditioning stimulation, while the presynaptic volleys evoked by test stimulation in WM remained unchanged, indicating that the effects of test stimulation remained unchanged before and after conditioning stimulation: therefore the potentiation of FPs can be ascribed entirely to enhanced synaptic transmission in VC. Fig. 4 B plots FPs as function of time intervals after the termination of conditioning stimulation and indicates a slight depression, a recovery toward the end of conditioning stimulation, and a rather gradual increase, which continues at least for two hours after the end of conditioning stimulation.

Corresponding to potentiation in the FPs, there was shortening in the orthodromic latencies of cortical units. In this particular case the FPs were rather strongly depressed during conditioning stimulation

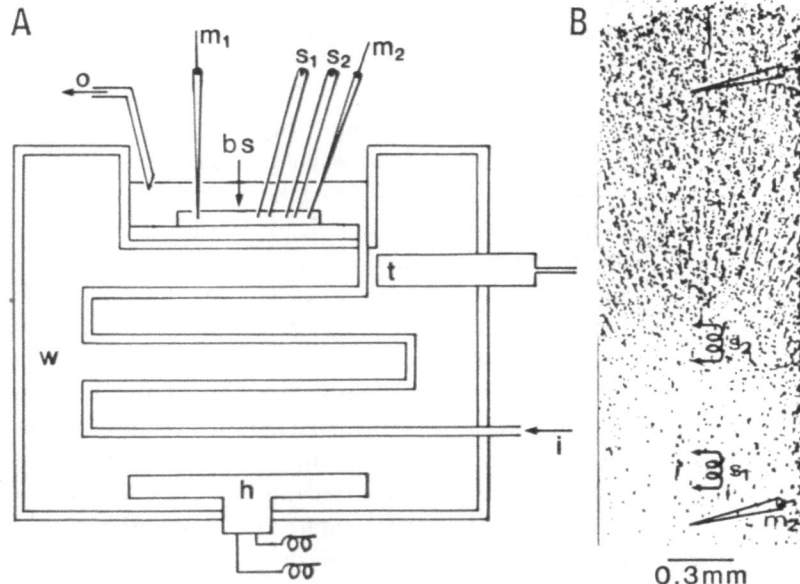

Fig. 3. Experimental arrangement for slice preparations. A) schematic diagram illustrating a recording chamber. *bs* brain slice. *m₁* and *m₂* glass microelectrodes. *S₁* and *S₂* stimulating electrodes. B) arrangements of electrodes illustrated in a histological selection of cortical slice

0.3mm

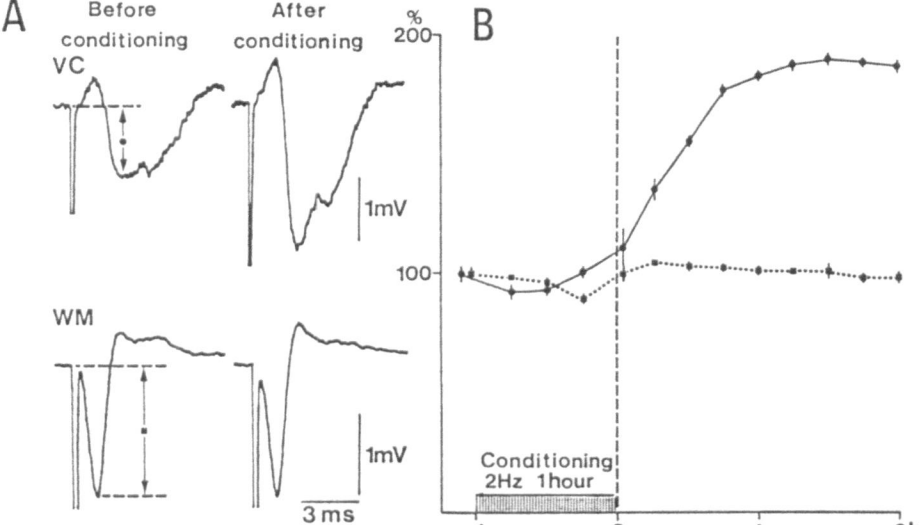

Fig. 4. Potentiation of field potentials evoked in area 17 of a slice preparation of infant kittens. A) field potentials (upper two traces) recorded from the visual cortex (VC) and the white matter (lower two traces) evoked by test stimulation before and after conditioning stimulation. B) changes in amplitude of the field potentials (FPs) in VC (filled circles) and WM (filled squares)

(Fig. 5 A). Orthodromic responses were completely suppressed during the period of conditioning stimulation. They recovered about 10 min after the end of conditioning stimulation with much larger latencies (6.5 ms) than before conditioning stimulation (5.6 ms). Afterwards the latency continued to decrease down to 4.6 msec (Fig. 5 B).

The shortening of the orthodromic latencies was systematically analyzed by sampling many cortical units in a single slice before and after conditioning stimulation. In this case test stimulation was applied at

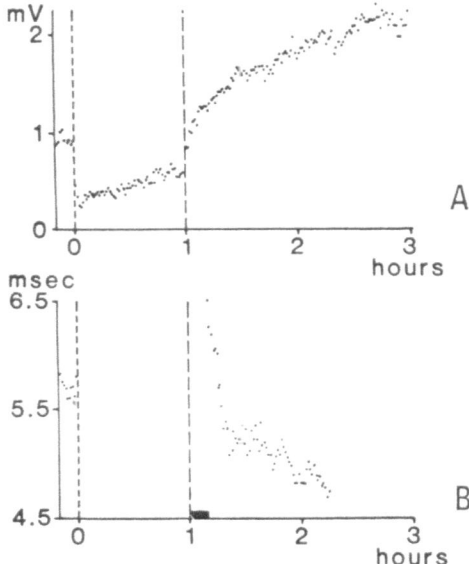

Fig. 5. Shortening of orthodromic latencies. A) similar plots of FPs to Fig. 4 B. B) similar to A) but for latencies of orthodromic responses in a cortical cell

two different sites in WM and the latencies were divided into a central delay for synaptic transmission and the conduction time of impulses. The histogram representing the central delays before conditioning stimulation was composed of two groups, one falling in a range of monosynaptic delay and the other in that of polysynaptic delay. Likewise, those after conditioning consisted of similar two groups. However, a tendency was noticed that both mono- and polysynaptic latencies were shorter after conditioning stimulation. The mean for monosynaptic delays was 0.8 msec before and 0.6 msec after conditioning, and that for polysynaptic delays was 2.2 and 1.4 msec. There was a similar difference in the lumped histograms of the central delays for the cortical cells sampled from 10 slices (Figs. 6 C and D): greater shortening in the polysynaptic latency (0.6 msec, from 2.3 to 1.7 ms) than monosynaptic latency (0.2 msec, from 0.9 to 0.7 ms), indicating that the potentiation is stronger at polysynaptic than monosynaptic transmission.

2. CD Analysis in Slice Preparations

The location of the synapses involved in the potentiation was studied by CD analysis. Local currents generating the FPs were determined as the second-order spatial differentials of the FPs, which were recorded as functions of the cortical depth. Fig. 7 illustrates the FPs and the local currents demonstrated by the CD analysis. Positivity in the current traces represents current-sinks, which is mostly ascribable to excitatory postsynaptic currents. Before conditioning stimulation the early currents (hatched area in the trace CD of Fig. 7 A), probably representing monosynaptic

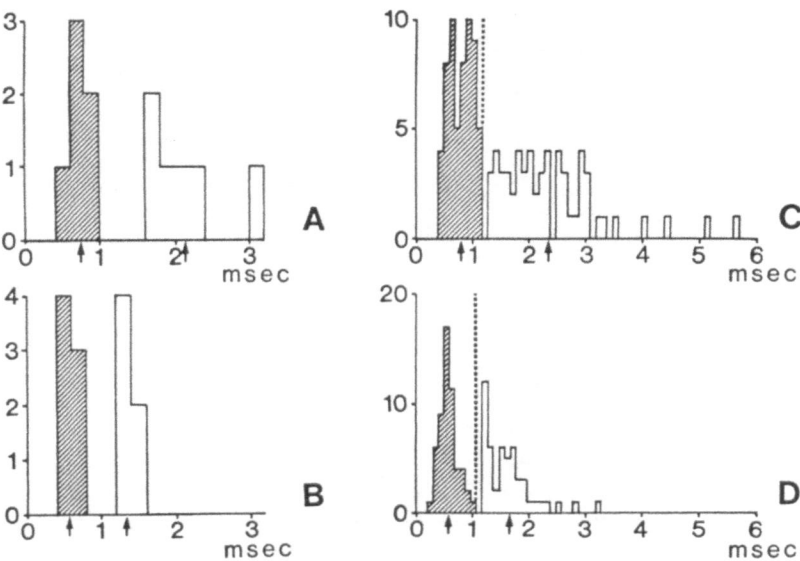

Fig. 6. Analysis of central delays. A) and B) central delays in cortical cells sampled from a single slice. A) before conditioning. B) after conditioning. C) and D) similar to A) and B), but for cortical cells sampled from 10 slices. Hatched columns, monosynaptic delays. Blank columns, polysynaptic delays

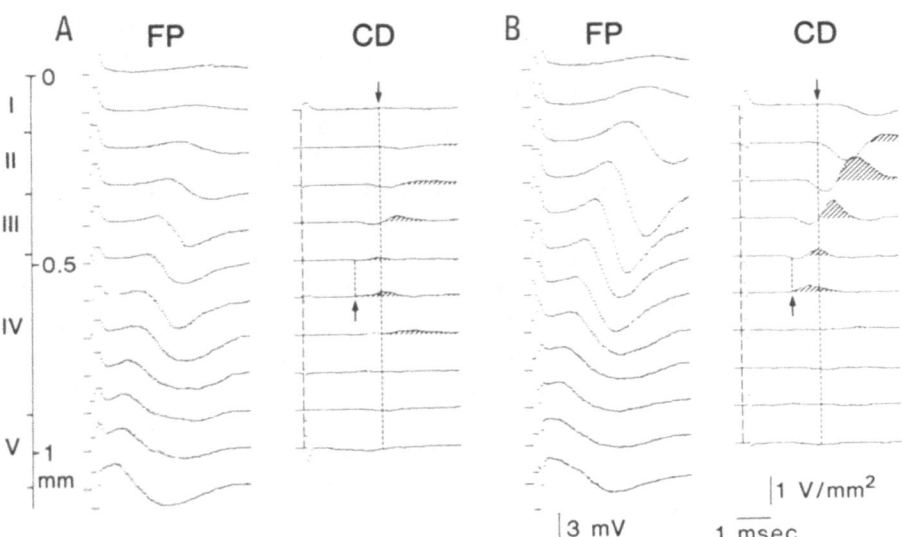

Fig. 7. CD analysis of cortical field potentials in area 17 of a slice preparation. A) and B) FPs (column *FP*) and local currents determined by CD analysis (*CD*) evoked in kitten VC by test WM stimulation. A) before conditioning stimulation. B) three hours after conditioning. Upward deflection represents a current-sink, and shading indicates major current-sinks. Upward and downward arrows indicate early and late currents (see text)

transmission occurred in layer IV, and the late currents, probably representing polysynaptic transmission in layers II and III. Three hours after conditioning stimulation both early and late currents were potentiated (Fig. 7 B). The potentiation was most pronounced in the late currents in layers II and III, and rather small in the early currents in layer IV, which agreed with the results of latency analysis that polysynaptic transmission was more strongly potentiated than monosynaptic transmission.

Distribution of the current-sinks in the cortical layers is represented by plotting the maximum amplitude of the early and late currents as functions of

the cortical depth. Before conditioning stimulation, the early currents (filled circles in Fig. 8 A) were localized in layer IV, while the late currents were predominantly distributed in layers II and III (filled circles in Fig. 8 B). Conditioning stimulation produced a 2-fold increase in the early currents in the superficial half of layer IV, and a 6-fold increase in the late currents throughout layers II and III.

Similar analysis made at different time intervals (1 and 11 hours) after conditioning stimulation indicated that the time course of the potentiation is different between the early and late currents. The early currents were almost double the control one hour after con-

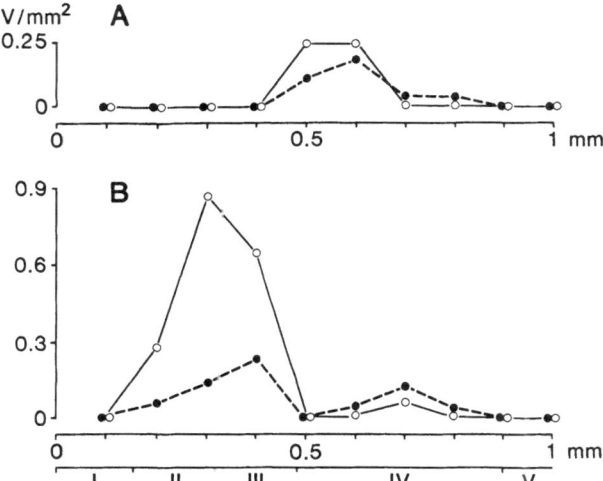

Fig. 8. Effects of conditioning stimulation on depth-profile of early and late currents. A) and B) depth profile of early and late currents illustrated in Fig. 1. Closed and open circles represent amplitude of the current-sinks before, and three hours after conditioning stimulation

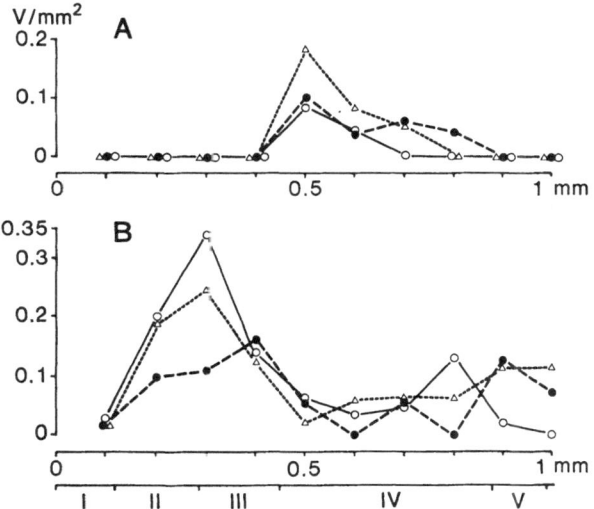

Fig. 9. Depth-profile of early and late currents at different time intervals after conditioning stimulation. A) and B) similar depth profiles to those in Fig. 8 at different time intervals (filled circles, control; open triangles, one hour; open circles 12 hours) after conditioning stimulation

ditioning stimulation, but returned almost to the control level at 12 hours (Fig. 9 A). In contrast, the late currents were double at one hour and continued to increase further up to threefold at 12 hours (Fig. 9 B).

The difference in the time course of the potentiation was further confirmed by plotting the maximal values of the early and late currents as functions of the time interval after conditioning stimulation. Potentiation of the early currents occurred even during the period of

conditioning stimulation, reached a maximum one hour after conditioning stimulation, and declined gradually for the succeeding 11 hours (circles in Fig. 10). In contrast, the late currents which were markedly depressed during conditioning stimulation, rapidly increased during the one hour following conditioning stimulation, and continued to increase rather gradually for the succeeding 11 hours (triangles).

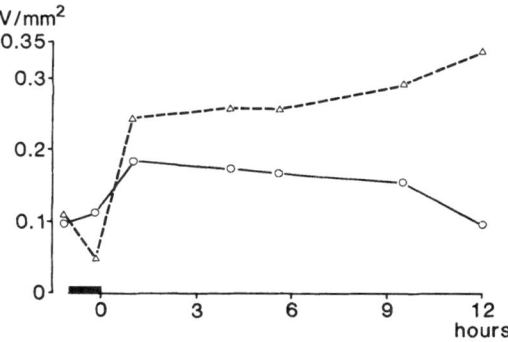

Fig. 10. Time-courses of changes in early and late currents after conditioning stimulation. The peak amplitudes of the early (open circles) and late currents (open triangles) are plotted as function of time intervals after conditioning

In similar studies made in 76 slices dissected from 59 infant kittens aged 7–49 days, it was found that potentiation of FPs was critically age-dependent. Potentiation occurred most frequently around the fourth week, in more than 80% of the slices at 21–34 days, moderately (50%) between 14–20 or 35–41 days, and never at earlier or later ages, between 7–13 and 42–49 days.

CD analysis was made in 11 slices dissected from kittens about four weeks old. Potentiation of the early currents was always less prominent in amplitude as well as in frequency than that of the late currents (Table 1). There were 80% potentiation in the early currents in layer IV and 180% potentiation in the late currents in layers II. Potentiation occurred in 9/11 for the early currents and in all cases for the late currents.

In summary, slice experiments demonstrated existence of synaptic plasticity in VC of the infant kittens. Both latency analysis of orthodromic unit responses, and CD analysis of FPs indicate that conditioning stimulation of WM produced long-lasting potentiation of synaptic transmission, which occurred more strongly but more slowly in polysynaptic transmission in the supragranular layer than monosynaptic transmission in the granular layer. However, this experimental paradigm has some ambiguity concerning

Table 1. *Results of CD Analysis of LTP Evoked in a Slice of Area 17 by White Matter Stimulation*

Layer	Early currents		Late currents	
	Enhancement	Shortening of latency (ms)	Enhancement	Shortening of latency (ms)
II	—	—	2.8 + 1.5 (11/11)	1.0 + 1.0 (4.0 + 1.0)
III	—	—	1.9 + 0.5 (11/11)	0.4 + 0.3 (2.5 + 0.5)
IVs	1.8 + 0.6 (9/11)	0.0 + 0.1* (1.3 + 0.3)	—	—
IVl	1.5 + 0.3 (5/11)	0.0 + 0.1* (1.4 + 0.4)	—	—
V	—	—	1.3 + 0.2 (5/11)	0.3 + 0.4 (3.1 + 0.4)

Values for enhancement were determined for 11 slices that exhibited LTP (see text). The ratio of the number of slices exhibiting clear enhancement to the total number of slices is indicated in brackets. Values for shortening were determined for these slices as the difference between current-sink latencies before and after conditioning stimulation. Figures in brackets represent the latency before conditioning stimulation. Minus sign indicates that current-sink were undetectable or too small to be analysed. * Indicates that current sink enhancement or shortening of latency is not statistically significant ($P > 0.05$ by Student's t test).

the pathway that triggers the plastic changes. Since WM stimulation excites various neuronal structures other than the afferent visual pathway, they might have been involved in this potentiation.

3. CD Analysis in Whole Animal Preparations

This possibility was excluded in similar experiments using LGN and optic chiasm (OC) stimulation in the whole animal preparations of infant kittens. The results were essentially similar. Fig. 11 illustrates FPs before and one hour after conditioning stimulation of the LGN. The CD analysis demonstrated similar early and late currents to those demonstrated by WM stimulation in slice preparations. There was a twofold increase in the early currents at layer IV, and a sixfold increase in the late currents in layers II and III.

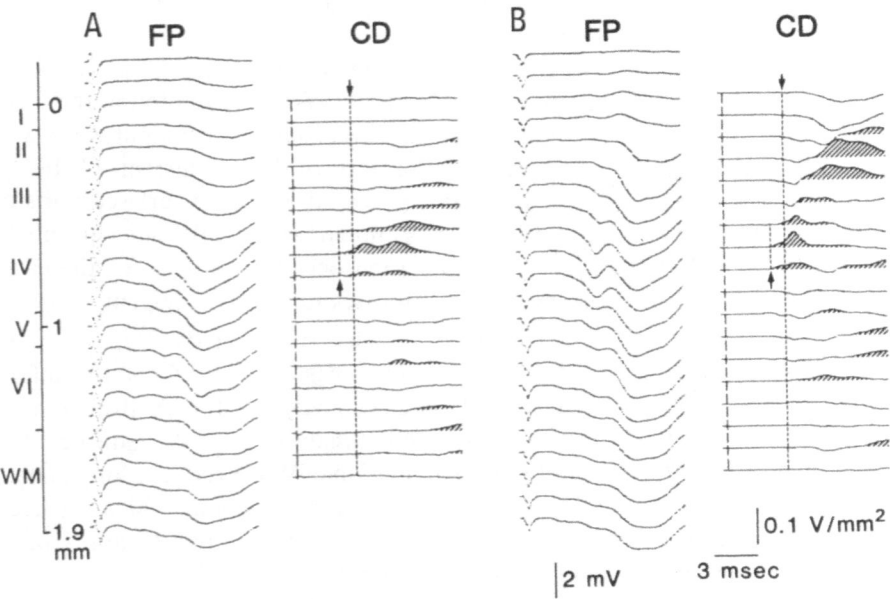

Fig. 11. CD analysis in area 17 of a whole animal preparation. A) and B) similar to Figs. 7 A and B

Table 2. *Results of CD Analysis of LTP in Area 17 Evoked by LGN Stimulation*

Layer	Early currents		Late currents	
	Enhancement	Shortening of latency (ms)	Enhancement	Shortening of latency (ms)
II	—	—	3.2 + 2.5 (8/8)	2.4 + 1.1 (8.8 + 1.2)
III	—	—	1.8 + 0.6 (8/8)	1.3 + 0.8 (5.8 + 0.6)
IVs	1.8 + 1.4 (6/8)	0.1 + 0.2* (2.6 + 0.5)	—	—
IVl	1.3 + 0.2 (2/8)	0.1 + 0.1* (3.1 + 0.6)	—	—
V	—	—	1.4 + 0.6 (5!/8)	0.8 + 1.0* (6.3 + 1.5)
VI	0.9 + 0.2 (1/8)	0.1 + 0.3* (2.6 + 0.5)	—	—

The parameters of the early and late currents were determined as in Table 1. 5! Includes two cases of undetectable control current-sink.

In nine similar experiments made in area 17 of the infant kittens at the ages around the fourth week, there was 3.2- and 1.8-fold potentiation in the late currents in layers II and III, and 1.8-fold potentiation in the early currents in layer IV (Table 2).

Similar results were obtained with OC stimulation in the whole animal preparations. One hour after conditioning stimulation, FPs were increased about twofold. Correspondingly, there was a twofold increase in the early currents in superficial half of layer IV, and a threefold increase in the late currents in layers II and III (Fig. 12 and Table 3).

Discussion

In summary, the effects of activation of the visual pathway were consistent between three types of experiments: with WM stimulation in the slice preparation

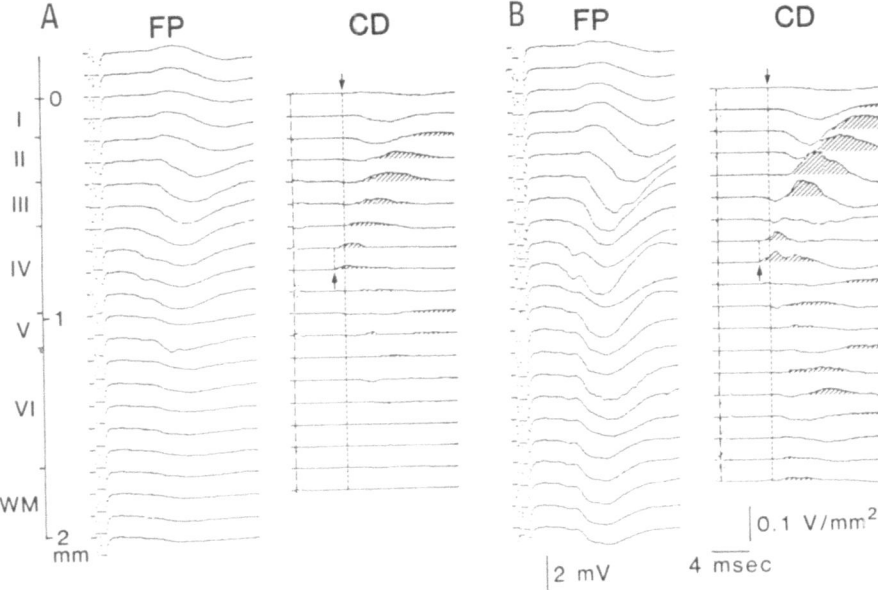

Fig. 12. CD analysis in area 17 of a whole animal preparation with OC stimulation. Similar to Figs. 7 and 11 but with OC stimulation

Table 3. *Results of CD Analysis of LTP in Areas 17 and 18 Evoked by OC Stimulation*

Layer	Early currents		Late currents	
	Enhancement	Shortening of latency (ms)	Enhancement	Shortening of latency (ms)
II	—	—	4.2 + 1.6 (4/4)	2.5 + 0.4 (13.3 + 2.4)
III	—	—	2.6 + 0.5 (4/4)	1.4 + 1.5* (9.2 + 3.3)
IVs	1.2 + 0.6* (2/4)	0.2 + 0.5* (4.8 + 1.1)	—	—
IVl	0.7 (0/4)	0.1 (4.0)	—	—
V	—	—	1.8 + 0.7* (3/4)	1.0 + 0.7* (9.7 + 1.9)
VI	1.6 + 0.8* (1/4)	0.0 + 0.1* (5.2 + 0.9)	—	—

The parameters of the early and late currents were determined as in Tables 1 and 2.

and with LGN and OC stimulation in the whole animal preparations. The results are: 1) activation of the visual pathway potentiated synaptic transmission at both geniculo-cortical and cortico-cortical synapses. 2) Potentiation was greater at the cortico-cortical than the geniculo-cortical synapses. 3) Potentiation occurred more slowly but lasted longer in the cortico-cortical than the geniculo-cortical synapses.

The long-lasting nature and the age-dependency of the potentiation demonstrated in the present experiments suggest that it may have a common basis with

modifiability of neuronal responsiveness by visual experience. In support of this view, similar changes in synaptic transmission to those produced by electrical stimulation of the visual pathway were found during normal development (Komatsu et al., 1985). In CD analysis made on FPs in VC of kittens at four different ages, LGN stimulation produced only early currents and no late currents in three-day-old kittens. The late currents developed between the ages of 19 and 33 days, in exactly the same way as they were produced after electrical stimulation of visual pathway, indicating that

AREA 17

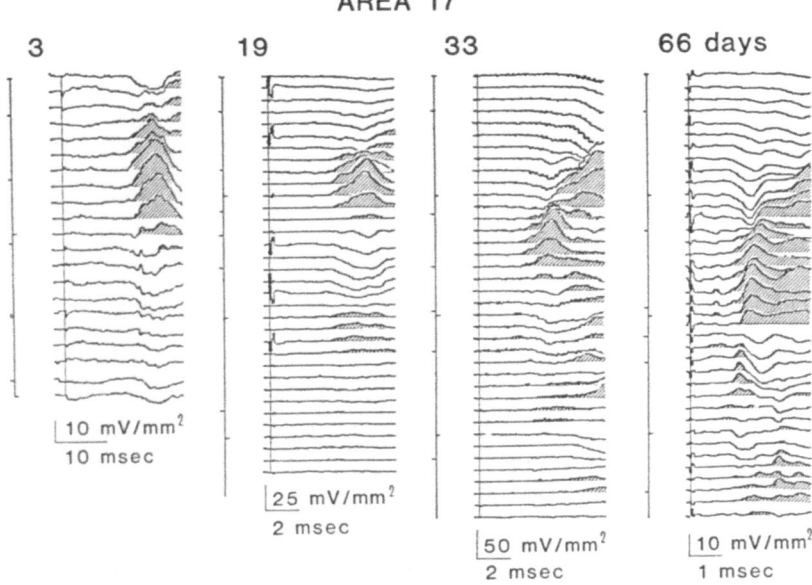

Fig. 13. Developement of visual cortical circuitry. The results of CD analysis made on FPs evoked by LGN stimulation in area 17 of whole animal preparations of kittens at four different ages are illustrated. (From Komatsu et al., 1985)

cortico-cortical synapses develop during that period (Fig. 13). Therefore, plastic changes are larger in the cortico-cortical synapses than the geniculo-cortical synapses after experimental activation of the visual pathway as well as in normal development. Photic responsiveness may be learned in highly plastic cortico-cortical synapses, while less plastic geniculo-cortical synapses may impose constraints for the learning which represent the prenatal design to control the learning. Layer II and III cells with cortico-cortical synapses are analogous to students learning visual function, and layer IV cells with geniculo-cortical synapses are teachers instructing the students what is to be learned. Thus, our answer to the Nature versus Nurture question is: neuronal responsiveness is nurtured by visual experience but it is under the constraints of the prenatal design.

References

1. Blakemore C, Cooper GF (1970) Development of the brain depends on the visual environment. Nature London 228: 477–478
2. Blakemore C, Mitchell DE (1973) Environmental modification of the visual cortex and the neural basis of learning and memory. Nature (London) 241: 467–468
3. Blakemore C, van Sluyters RC (1975) Innate and environmental factors in the development of the kitten's visual cortex. J Physiol (London) 248: 663–716
4. Blasdel GG, Mitchell DE, Muir DW, Pettigrew JD (1977) A physiological and behavioural study in cats of the effect of early visual experience with contours of a single orientation. J Physiol (London) 265: 615–636
5. Buisseret P, Imbert M (1976) Visual cortical cells: Their developmental properties in normal and dark reared kittens. J Physiol (London) 255: 511–525
6. Hebb DO (1949) The organization of behaviour. Wiley & Sons, New York
7. Hirsch HVB, Spinelli DN (1970) Visual experience modifies distribution of horizontally and vertically oriented receptive fields in cats. Science 168: 869–871
8. Hubel DH, Wiesel TN (1963) Receptive fields of cells in striate cortex of very young, visually inexperienced kittens. J Neurophysiol 26: 994–1002
9. Hubel DH, Wiesel TN (1970) The period of susceptibility to the physiological effects of unilateral eye closure in kittens. J Physiol (London) 206: 419–436
10. Komatsu Y, Fujii K, Nakajima S, Umetani K, Toyama K (1985) Electrophysiological and morphological correlates in the development of visual cortical circuity in infant kittens. Dev Brain Res 22: 305–309
11. Komatsu Y, Toyama K, Maeda J, Sakaguchi H (1981) Long-term potentiation investigated in a slice preparation of striate cortex of young kittens. Neurosci Letters 26: 269–274
12. Mitzdorf U, Singer W (1978) Prominent excitatory pathway in cat visual cortex (A 17 and A 18): a current source-density analysis of electrically evoked potentials. Exp Brain Res 33: 371–394
13. Rauschecker JP, Singer W (1981) The effects of early visual experience on the cat's visual cortex and their possible explanation by Hebb synapses. J Physiol (London) 310: 215–239
14. Singer W, Freeman B, Rauschecker J (1981) Restriction of visual experience to a single orientation affects the organization of orientation columns in cat visual cortex: a study with deoxyglucose. Exp Brain Res 41: 199–215
15. Stryker MP, Sherk H, Leventhal AG, Hirsch HVB (1978) Physiological consequences for the cat's visual cortex of effectively restricting early visual cortex with oriented contours. J Neurophysiol 41: 896–909
16. Toyama K, Matsunami K, Ohno T, Tokashiki S (1974) An intracellular study of neuronal organization in the visual cortex. Exp Brain Res 21: 45–66
17. Wiesel TN, Hubel DH (1963) Single cell responses in striate cortex of kittens deprived of vision in one eye. J Neurophysiol 26: 1003–1017
18. Wiesel TN, Hubel DH (1965) Comparison of the effects of unilateral and bilateral eye closure on cortical unit responses in kittens. J Neurophysiol 28: 1029–1040

Correspondence: Keisuke Toyama, M.D., Department of Physiology, Kyoto Prefectural School of Medicine, Kawaramachi, Kamikyo-ku, Kyoto 602, Japan.

Acta Neurochirurgica, Suppl. 41, 78–84 (1987)

Morphological Aspects of Formation of Neuronal Pathways in the Chick Spinal Cord—Golgi and Electron Microscopic Studies

A. Kanemitsu, S. Matsuda, and **Y. Kobayashi**

Department of Neuroanatomy, Institute of Brain Research, School of Medicine, University of Tokyo, Tokyo, Japan

Summary

The early formation of neuronal connections in the cervical cord of chick embryos at stages 17 to 31 was studied by observing axonal courses with Golgi preparations and the distribution of synapses with electron microscopy. The results are summarized as follow:

1. Synaptic contacts between spinal interneurons and ipsilateral as well as controlateral motor neurons first develop at stage 22. The early central pathways from the dorsal root to the ventral root may be formed at stage 25 with intervention of interneurons of the primordial dorsal horn, which is composed of neurons of the zona spongiosa and the nucleus proprius of the dorsal horn, and at stage 27 with the intervention of interneurons of the zona intermedia, the nucleus proprius of the ventral horn and the primordial dorsal horn as well.

2. These polysynaptic, bilateral central pathways appear to be established just before the arrival of supraspinal descending fibres at the cervical cord, and one or two days before the formation of ipsilateral monosynaptic spinal reflex arch.

3. These early spinal central pathways are connected by synapses with spherical synaptic vesicles, and almost all of these synapses are of the axo-dendritic type and located in the spinal white matter.

Keywords: Synaptogenesis; hodogenesis; cervical cord; chick embryo; TEM; Golgi method.

In order to examine early developing spinal central pathways, we studied in the embryonic chick cord 1.) the course of axons and the growth of dendrites of early differentiated spinal neurons and spinal ganglion cells by the Golgi method and 2.) the distribution of synapses by making electron microscopic photomontages. The findings obtained by these morphological techniques suggest that the ipsi- and contralateral, polysynaptic central pathways from the dorsal root to the ventral root are the first spinal neuronal pathways to develop.

Materials and Methods

For Golgi studies, the upper cervical parts of embryos at stages 24 (4.5 days) to 41 (15 days) were treated for Golgi preparations according to the modified Golgi method by Valverde[17], and cut into transverse or sagittal serial sections of 100 µm thickness. The embryos were stage according to the stage table by Hamburger and Hamilton[3]. For each staged, the whole area of the cervical cord of ten successive sections was photographed at a magnification of × 100, and ten photomontages were made at a final magnification of × 270. Impregnated dendrites, cell bodies, axons and axon collaterals were drawn on a tracing paper covering the photomontage.

For electron microscopic studies, the upper cervical cord was dissected at stages 17 (2.5 days), 18 (3 days), 22 (3.5–4 days), 25 (4–4.5 days), 27 (5–5.5 days), 29 (6–6.5 days) and 31 (7–7.5 days) in a solution of 4% paraformaldehyde and 1% glutaraldehyde in 0.1 M phosphate buffer, and fixed in a solution of 3% glutaraldehyde in the same buffer for one hour. Blocks were then treated by conventional techniques for electron microscopy. For each stage, the entire one-half side of the cervical cord was photographed at a magnification of × 2000. The entire half side was covered by 36, 48, 68, 94, 153, 165 and 181 negatives for each stage. Seven electron microscopic photomontages were made at a final magnification of × 3500.

Results and Discussion

Early Differentiated Spinal Neurons and Ganglion Cells

In our previous ^{3}H-thymidine autoradiographic studies[5, 6] we determined the time of origin (time of the final DNA synthesis) of chick spinal neurons and ganglion cells. Early developing spinal neuronal pathways are formed by early differentiated spinal neurons and ganglion cells, so, in advance we will mention what spinal neurons and ganglion cells differentiate early. In the ^{3}H-thymidine autoradiograms, heavily labelled neurons are interpreted as neurons whose final DNA synthesis occurred just at the stage of injection of isotope[13]. Therefore, it is clear from the distribution of heavily labelled neurons shown in Fig. 1 that neurons of the motor nucleus (h), the zona intermedia (f), the zona spongiosa (a) and ganglion cells of the ventrolateral part of the spinal ganglion (SG) are the first to differentiate, followed by neurons of the nucleus proprius of the ventral horn (g) and of the dorsal horn (c) and also ganglion cells of the dorsomedial part of the spinal ganglion, and the last to originate are neurons of the substantia gelatinosa (b) and the neck (d) and the base (e) of the dorsal horn.

Each cytoarchitectonic division of the spinal grey matter is composed of more or less heterogenous neurons in respect to their time of origin, so that heavily labelled neurons were counted in each division in each case, to determine the order of development among these divisions. The curves thus obtained indicate the number of neurons which differentiated at each stage for each division (Fig. 2). It is well known that motor neurons differentiate very early, but this does not mean

that the differentiation of motor neurons precedes that of spinal interneurons and ganglion cells. As is obvious in Fig. 2, motor neurons differentiate during stages 14 to 20, whereas during these stages, a fairly large number of ganglion cells and interneurons of the zona intermedia and a small number of interneurons of the zona spongiosa, the nucleus proprius of the dorsal horn and of the ventral horn also differentiate.

The Course of Axons of Spinal Interneurons and Ganglion Cells

The Bodian sections prepared from the cervical cord of the embryo at stage 20 (3–3.5 days) showed that the proximal processes of young spinal ganglion cells already reached the dorsal funiculus through the dorsal root and the axons of motor neurons grew out of the neural tube to form the ventral root, while all of the

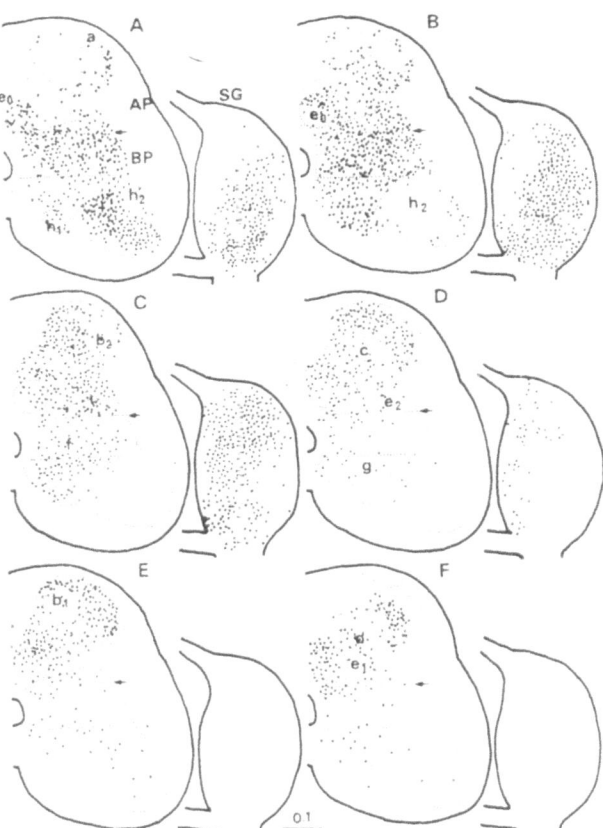

Fig. 1. Distribution of the spinal neurons and the ganglion cells differentiated at stages 17 (A), 20 (B), 23 (C), 26 (D), 28 (E), and 30 (F) in the brachial cord[5]. Heavily labelled neurons of ten successive sections were superimposed in each of the experimental cases, in which the embryos received 20 µci of ^{3}H-thymidine at stages 17, 20, 23, 26, 28, and 30, respectively, and they were all killed at stage 36. The arrows indicate the boundary between the alar plate (AP) and the basal plate (BP). a to h cytoarchitectonic divisions of the spinal grey matter (see the text). SG Spinal ganglion; e_o columna dorsalis magnocellularis of Huber (1936), a nucleus characteristic of the avian spinal cord. Unit of the scale bar = mm

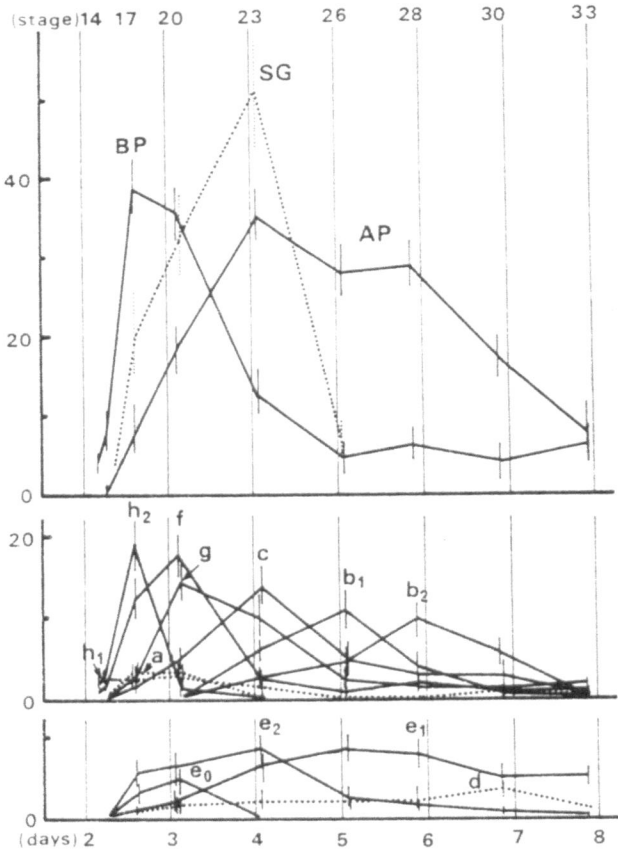

Fig. 2. Change in the number of differentiated spinal neurons and ganglion cells in each cytoarchitectonic division according to the developmental stages[5]. Each curve respresents the mean number of the neurons (ordinate) for a 10 µm thick section, which differentiated at each stage of injection of isotope (abscissa) in the basal plate (BP), the alar plate (AP), the spinal ganglion (SG) and each divison (a–h) of the brachial cord. The vertical lines indicate the 95% confidence interval for the true mean

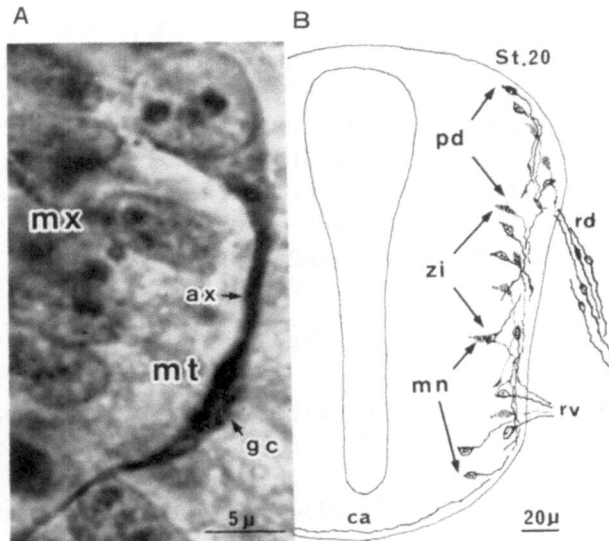

Fig. 3. A young spinal interneuron (A) and the course of axon of spinal neurons and ganglion cells (B) in the cervical cord of the chick embryo at stage 20. Neurons were photographed and traced from the transverse sections impregnated with the Bodian method. *mx* Matrix layer; *mt* mantle layer; *ax* axon; *gc* growth cone; *pd* neurons of the primordial dorsal horn, which is composed at this stage of neurons of the zona spongiosa (cf. Figs. 1 and 2); *zi* neurons of the zona intermedia; *mn* motor neurons; *ca* white ventral commissure; *rd* dorsal root, *rv* ventral root

early differentiated spinal interneurons sent their axon ventralwards (Fig. 3). Since these spinal interneurons differentiated before stage 20, they can be regarded as those of the zona spongiosa and the zona intermedia, according to our findings obtained by ^{3}H-thymidine autoradiography (cf. Figs. 1 and 2). As the stage advanced, these ventralward growing axons of spinal interneurons reached the ipsilateral ventral or lateral funiculus, or the contralateral ventral funiculus (Fig. 4, ST 24, ST 27). More than 60% of impregnated interneurons were commissural. After reaching the ventral or the lateral funiculus, these stem axons bifurcated into an ascending and a descending funicular axon (Fig. 4, ST 28). The proximal processes of spinal ganglion cells also bifurcate in the ipsilateral dorsal funiculus in the same fashion as the axon of the spinal interneurons[11, 16]. Therefore, in the early stages of development, the dorsal funiculus and the ventral and the lateral funiculus are composed exclusively of axons of spinal ganglion cells and of spinal interneurons, respectively.

As for the axon collaterals growing from the funicular axons into the spinal grey matter, they were not found inside the grey matter at stage 24, while at stage 27, they were observed invading from the ventral, the lateral and the dorsal funiculus as well (cf. Fig. 4, ST 27 B, left half). It was already suggested by Cajal[11] and also by Windle and Orr[16] that it is axon collaterals or secondary collaterals growing from the former that make synaptic contacts with spinal neurons (Fig. 5).

The Distribution of Synapses and the Growth of Dendrites

In the electron microscopic study, the thickening of apposed membranes and the presence of a few synaptic vesicles are regarded as two major morphological signs for a furture synapse[2, 14, 15, 18]. According to these criteria we examined the distribution of synapses in the micro-photomontages. At stages 17 and 18, synaptic vesicle-like structures were sometimes found, but no membrane thickening could be observed there, so that they were not taken as positive findings. Synapses were first observed at stage 22 in the marginal layer near the exit of the ventral root (Fig. 6). At stage 25, they were found in the dorsal half of the dorsal funiculus, in the ventral half of the lateral funiculus and in the ventral funiculus, and at stage 27, they were observed in every place along the border of the mantle layer. At stages 29 and 31, synapses were distributed not only along the border of the spinal grey matter, but were also scattered to the peripheral part along the outer surface of the cord. These synapses were all of the axo-dendritic type. As for the shape of synaptic vesicles, they were all of the spherical type at stages 22 to 27, and at stage 29 flattened synaptic vesicles began to appear.

It should be noticed that the synapses observed at stages 22 to 27 were all located in the vicinity of the border of the spinal grey matter. This fact, together with the findings obtained by the Golgi method, suggests that the axon collaterals coming from the funiculi make synaptic contact chiefly with the dendrites growing into the funiculi, as proposed by Windle and Orr[16].

At stages 29 and 31, synapses were distributed all over the spinal white matter. This fact may be explained by the finding that after stage 27, many dendrites were observed to invade the white matter for a fairly long distance, and that some dendrites reached the outer surface of the cord[8] (Fig. 7 A, B). Synapses were found all over the spinal grey matter after stage 33[9]. Thus, the spinal white matter is the site of neuronal connections in the early embryonic stages. Accordingly, neurons located deep in the spinal grey matter have to send their

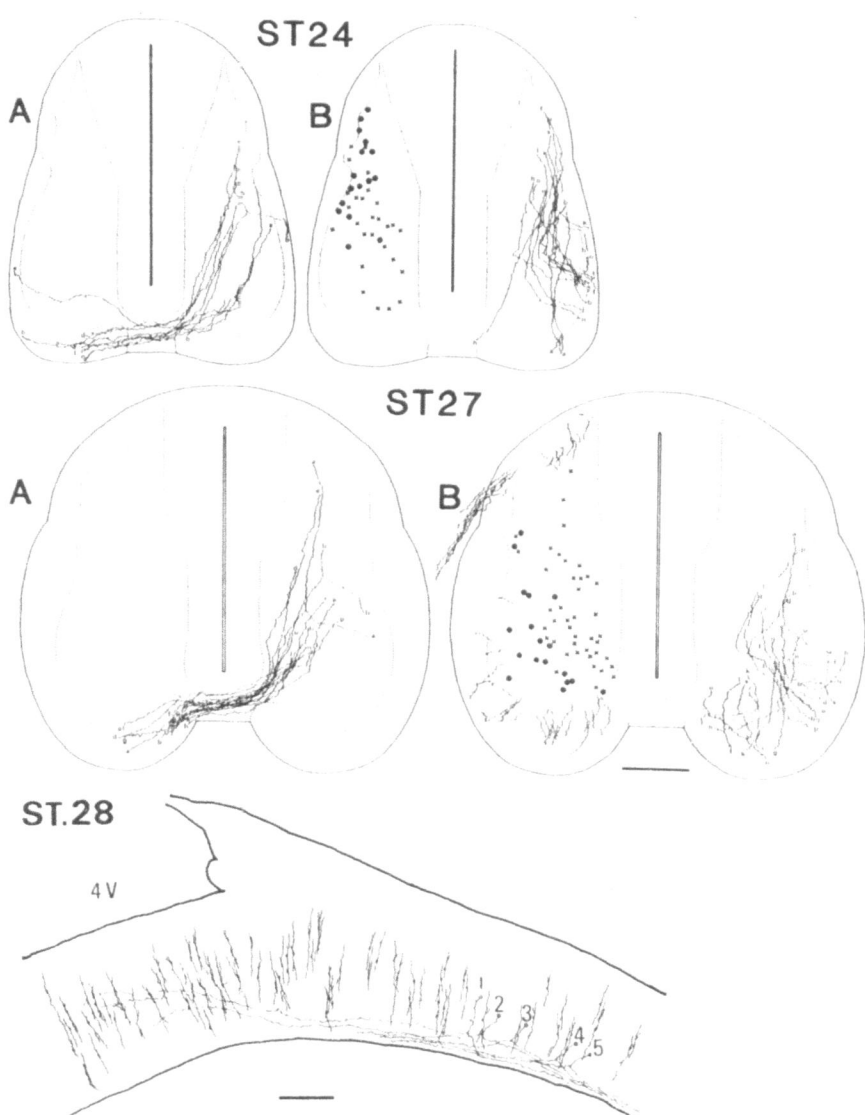

Fig. 4. Course of the axons of spinal interneurons in the transverse section (ST 24, ST 27)[7] and in the sagittal section (ST 28)[8]. The small letters in the grey matter and the white matter in figures ST 24 and ST 27 indicate the location of cell bodies and the cut end of corresponding axons, respectively. Axons were traced from ten successive transverse Golgi sections. Filled circles and crosses in the left half of ST 24 and ST 27 indicate the location of cell bodies of association interneurons (B) and commissural interneurons (A), respectively. Five open circles (*1–5*) in ST 28 indicate the location of cell bodies of spinal interneurons, from which the stem axon grows into the ventral funiculus to bifurcate there into an ascending and a descending funicular axon. Spinal interneurons (right side) and neurons of the reticular formation (left side) expand their dendrites perpendicularly to the neural axis. 4 V Fourth ventricle. Scale bar for ST 24, ST 27 = 0.1 mm, for ST 28 = 0.2 mm

dendrites to the white matter as quickly as possible to make synaptic contact early. Young interneurons usually had one long initial dendrite and one or two shorter dendrites growing out of the cell body later than the former (cf. Fig. 7 A, B). In our previous ³H-thymidine autoradiographic studies, we showed that early differentiated neurons were larger, and late differentiated ones smaller[4, 6]. So, at each stage we measured the length of the initial dendrite of impregnated larger interneurons, and calculated their growth rate. At stage 24, the initial dendrite of interneurons was less than 70 µm, and the growth rate was 92–184 µm per day at stage 24, 46–92 µm at stage 27, 15–22 µm at stages 29 to 33 and 4–5 µm after stage 36 (Fig. 8). The quick growth of the initial dendrite in the early stages might be interpreted as a highly effective mechanism for the spinal interneurons to make spinal neuronal pathways in the early embryonic stages.

Possible Initial Pathways from the Dorsal Root to the Ventral Root

According to the early Golgi studies by Cajal[11] and Retzius[12] the dendrites of motor neurons protrude into the ventrolateral part of the marginal layer, and the axons of early differentiated spinal interneurons already reach the ventral and the lateral funiculus in large numbers as early as the fourth day of incubation (stages 22 to 23). Therefore, the synapses we observed

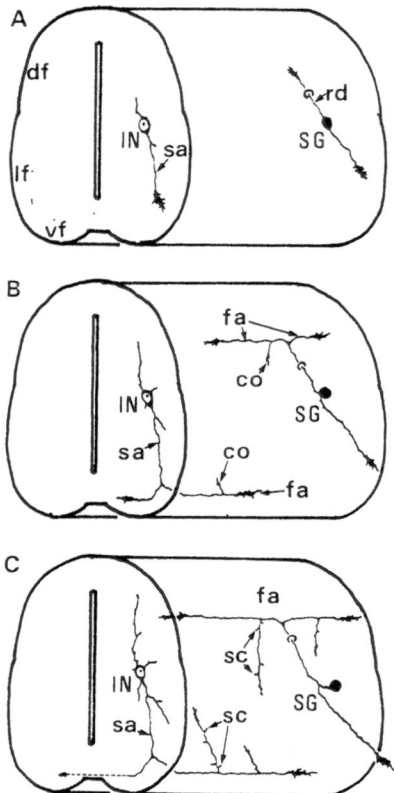

Fig. 5. Schema showing the course of axons of the spinal interneuron (*IN*) and the spinal ganglion cell (*SG*). The developmental stage advances from A to C. *df, lf, vf* dorsal, lateral, and ventral funiculus; *rd* dorsal root, *sa* stem axon; *fa* ascending and descending funicular axon; *co* axon collaterals growing from funicular axons into the spinal grey matter, *sc* secondary collaterals growing from axon collaterals to make synaptic contact

at stage 22 in the ventrolateral corner of the cervical cord could be interpreted as those connecting the axons or the axon collaterals of spinal interneurons with the dendrites of motor neurons. Thus, it is highly probable that the synaptic contacts between spinal interneurons and motor neurons are the first to develop.

Since those neurons which sent their axon to the dorsal funiculus could not be found at stage 24 nor at stage 27, the dorsal funiculus in early stages may well be composed exclusively of the axons of primary sensory neurons. Existence of synapses in the dorsal funiculus at stage 25 suggests the beginning of the formation of connections between dorsal root fibres and inter- neurons located in the primordial dorsal horn. These spinal interneurons in the primordial dorsal horn sent their axon by a long route to the ispilateral or the contralateral ventral funiculus or to the ipsilateral lateral funiculus (cf. Fig. 4). The primordial dorsal horn at this stage is composed of interneurons of the zona

spongiosa and the nucleus proprius of the dorsal horn (cf. Figs. 1 and 2). Thus, early neuronal pathways from the dorsal root to the ventral root may be formed at stage 25 with the intervention of interneurons of the zona spongiosa and the nucleus proprius of the dorsal horn. These pathways may be both ipsilateral and contralateral, for more than 60% of the impregnated spinal interneurons were commissural (cf. Fig. 4).

The appearance of synapses newly observed at stage 27 in the dorsal half of the lateral funiculus (cf. Fig. 6, ST 27) suggests the beginning of synaptic contact between the axons or the axon collaterals of the lateral funiculus and the dendrites of interneurons located in the zona intermedia. On the other hand, interneurons in the primordial dorsal horn sent their axon to the ipsilateral lateral funiculus (cf. Fig. 4). Therefore, syn- aptic connections between interneurons in the primor- dial dorsal horn and those in the zona intermedia are likely to be formed at this stage. Interneurons in the zona intermedia, in turn, sent their axon to the ipsi- lateral or the contralateral ventral funiculus (cf. Fig. 4). At this stage a fairly large number of interneurons of the nucleus proprius of the ventral horn have already differentiated (cf. Figs. 1 and 2). Thus, it is conceivable that at stage 27, dorsal root fibres and motor neurons are connected ipsilaterally and contralaterally with the intervention of interneurons of the zona intermedia, the nucleus proprius of the ventral horn and the primordial dorsal horn.

Long axon collaterals coming from the dorsal funiculus reached the ventral funiculus through the spinal grey matter at stage 30 (cf. Fig. 7 C). Since synapses were not found in the spinal grey matter before stage 33, it is unlikely that these long axon collaterals make synaptic contact with the dendrites of motor neurons growing dorsally to the zona inter- media. Thus, we can reasonably assume that the monosynaptic reflex arch may be formed after stage 30.

As for the supraspinal descending fibres, according to Okado and Oppenheim[10], a few neurons of the recitular formation and the vestibular nucleus are retrogradely labelled after the injection of HRP into the cervical cord of five-day chick embryos (stage 26), and a good number of neurons of these nuclei are labelled by the injection after six days of incubation (stages 28 to 29). Supraspinal descending fibres descend in the ventral and the lateral funiculus[1]. Therefore, the greater part of the synapses we observed in the ventral and the lateral funiculus at stage 27 can be regarded as formed by the axons or the axon collaterals of spinal inter- neurons, and not by supraspinal descending fibres.

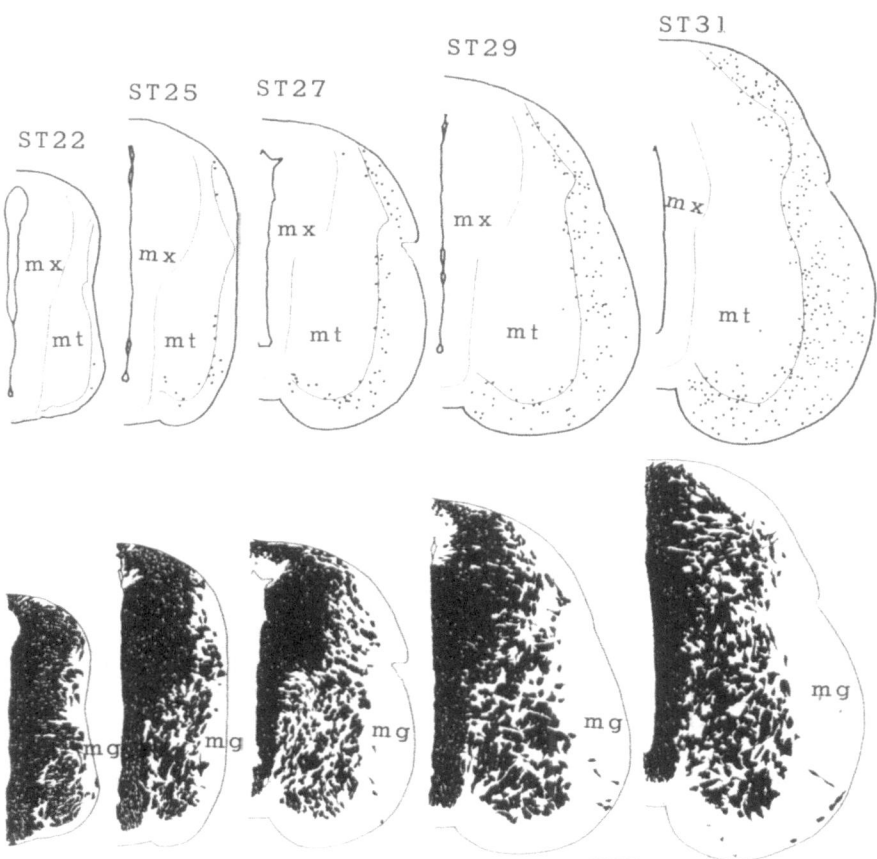

Fig. 6. Distribution of synapses (top) and matrix cells and spinal neurons (bottom). One dot represents one synapse. Synapses were 2, 19, 57, 107 and 237 at stages 22 to 31, respectively. Matrix cells and spinal neurons were drawn from the microphotomontages in which the distribution of synapses was studied. *mx, mt, mg* Matrix, mantle, and marginal layer. Scale bar = 0.1 mm

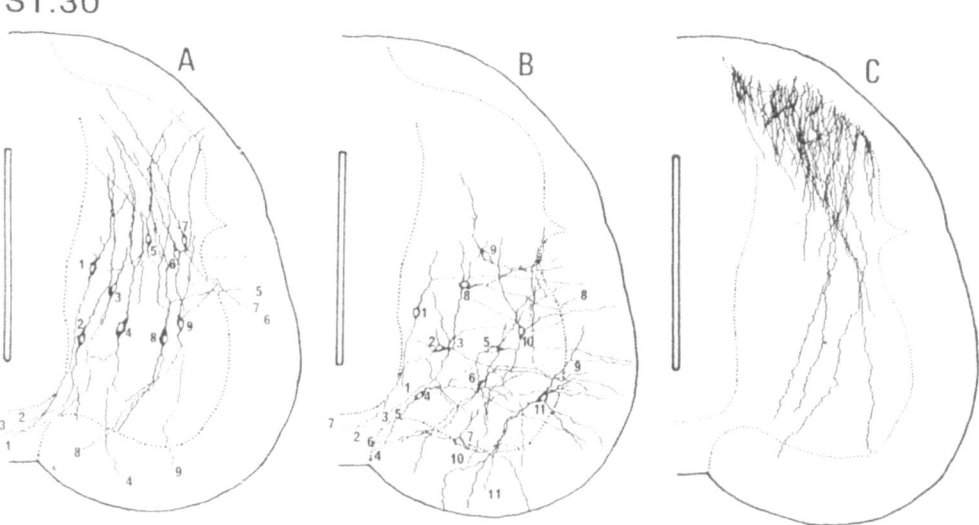

Fig. 7. Dendritic pattern of interneurons in the zona intermedia (*A, B*) and axon collaterals coming from the dorsal funiculus (*C*). Impregnated dendrites and axon collaterals were drawn from the Golgi preparations obtained from the upper cervical cord of the chick embryo at stage 30. Spinal interneurons extend their initial dendrite to the dorsal funiculus (*A*), the lateral and the ventral funiculus (*C*). Around this stage, dendrites are relatively long compared with the size of the spinal cord in the transverse section (see the explanation of Fig. 8). The arabic numerals indicate the location of cell bodies and the cut end of corresponding axons. Most of the axon collaterals coming from the dorsal funiculus end in the primordial dorsal horn; some can be followed to the zona intermedia and long collaterals reach the ventralmost part of the ventral horn at this stage. Scale bar = 0.1 mm

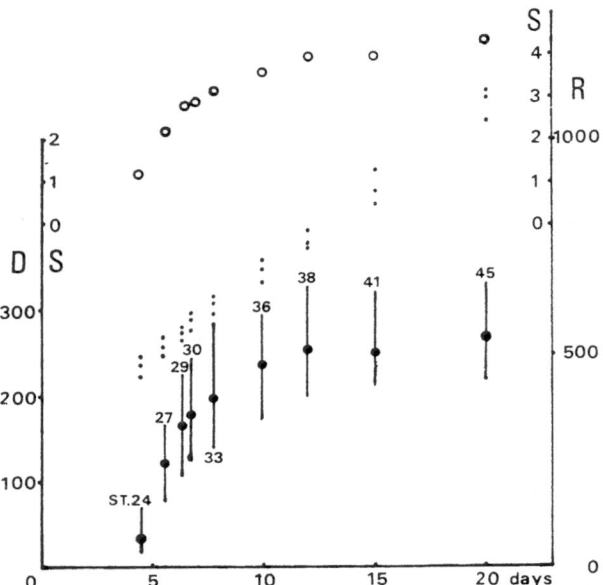

Fig. 8. Changes in the number of stem dendrites (top, *S*), the dorsoventral diameter of the cervical cord (middle, *R*) and the length of the initial (longest) dendrite (bottom, *D*) according to the developmental stages[8]. The dorsoventral diameter of the cervical cord was measured on three embryos for each stage. The vertical bars of the dendritic length represent the maximum and the minimum value of the initial dendrites, and the dots indicate their mean values. It should be noted that the growth of the initial dendrite slows down as the number of late developing stem dendrites increases with the age. The growth rate of the initial dendrite is greater than that of the cord diameter before stage 33, while this relationship becomes inverse thereafter. This would explain why dendrites look very long at stage 30 (cf. Fig. 7 A, B). Unit of the ordinates (D, R) = μm

References

1. Cohen DH (1974) The structural organization of avian brain: an overview. In: Goodman IJ, Schein MW (eds) Birds-brain and behaviour. Academic Press, New York, pp 20–73
2. Glees P, Sheppard BL (1964) Electron microscopical studies of the synapse in the developing chick spinal cord. Z Zellforsch 62: 356–362
3. Hamburger V, Hamilton HL (1951) A series of normal stages in the development of the chick embryo. J Morphol 88: 49–92
4. Kanemitsu A (1971) Relation entre la taille des neurones et leur époque d'apparition dans la moelle épinière chez le poulet—étude autoradiographique et caryométrique. Proc Japan Acad 47: 432–437
5. Kanemitsu A (1972) Étude quantitative de la neurogénèse dans la moelle épinière chez le poulet par l'autoradiographie. Proc Japan Acad 48: 758–763
6. Kanemitsu A (1979) Developmental sequence among spinal neurons. In: Tsubaki T, Toyokura Y (eds) Amyotrophic lateral sclerosis. Univ Tokyo Press, Tokyo, pp 387–403
7. Kanemitsu A, Matsuda S (1984) Synaptogenesis in the chick cervical cord and possible initial central pathways from dorsal root fibers to motor neurons—Golgi and electron microscopic studies. Neurosci Lett 48: 1–6
8. Kanemitsu A, Matsuda S (1985) Growth of the initial dendrite of early differentiated interneurons in the zona intermedia of the embryonic chick cervical cord—a morphometric study with Golgi method. Neurosci Lett 53: 267–272
9. Matsuda S, Kanemitsu A (1985) Distribution of synapses in the upper cervical cord of chick embryo from stage 28 to 38. Acta Anat Nippon 60: 361
10. Okado N, Oppenheim RW (1985) The onset and development of descending pathways to the spinal cord in the chick embryo. J Comp Neurol 232: 143–161
11. Ramón y Cajal (1890) A quelle époque apparaissent les expansions des cellules nerveuses de la moelle épinière du poulet? Anat Anz 5: 631–639
12. Retzius G (1893) Zur Kenntniss der ersten Entwicklung der nervösen Elemente im Rückenmarke des Hühnchens. Biologische Untersuchungen, Bd 5. Vogel, Leipzig, pp 48–58
13. Sidman RL (1970) Autoradiographic methods and principles for study of the nervous system with thymidine-H[3]. In: Nauta WJH, Ebbesson SOE (eds) Contemporary research methods in neuroanatomy. Springer, Berlin Heidelberg New York, pp 252–274
14. Skoff RP, Hamburger V (1974) Fine structure of dendritic and axonal growth cones in embryonic chick spinal cord. J Comp Neurol 153: 107–147
15. Stelzner DJ, Martin AH, Scott GL (1973) Early stages of synaptogenesis in the cervical spinal cord of the chick embryo. Z Zellforsch 138: 475–488
16. Windle WF, Orr DW (1934) The development of behaviour in chick embryo: spinal cord structure correlated with early somatic motility. J Comp Neurol 60: 287–306
17. Valverde F (1965) Studies on the piriform lobe. Harvard Univ Press, pp 4–9
18. Vaughn JE, Grieshaber JA (1973) A morphological investigation of an early reflex pathway in developing rat spinal cord. J Comp Neurol 148: 177–210

Correspondence: Dr. A. Kanemitsu, Department of Neuroanatomy, Institute of Brain Research, School of Medicine, University of Tokyo, Hongo 7-3-1, Bunkyo-ku, Tokyo 113, Japan.

Acta Neurochirurgica, Suppl. 41, 85–94 (1987)
© by Springer-Verlag 1987

A Novel Concept on the Pathogenetic Mechanism Underlying Ischaemic Brain Oedema: Relevance of Free Radicals and Eicosanoids

T. Asano, T. Shigeno, H. Johshita, M. Usui*, and T. Hanamura**

Department of Neurosurgery, Saitama Medical Centre, Saitama Medical School, *Division of Neurosurgery, Aizu Chuo Hospital, **Department of Neurosurgery, University of Tokyo Hospital, Tokyo, Japan

Summary

A survey on literature reports and our own experimental studies on the pathogenetic mechanisms underlying ischaemic brain oedema is given and a new concept proposed.

In regional incomplete ischaemia the lipoxygenase activity is enhanced, presumably caused by an increase of free radicals and hydroperoxides, leading to an enhancement of endothelial Na^+, K^+-ATPase and increased sodium and water transport from blood to brain.

The aggravation of brain oedema and post-ischaemic hypoperfusion following recirculation appears to be mainly due to an activation of the cyclo-oxygenase pathway with release of oxidants from PGG_2, which causes non-specific but detrimental damage to the endothelial and parenchymal cells.

This new concept may open future perspectives in treatment which are briefly discussed.

Keywords: Brain oedema; arachidonate cascade; free radical; lipoxygenase; cyclo-oxygenase; Na^+, K^+-ATPase; brain microvessels.

During the past decade, a considerable research effort has been directed towards substantiation of the hypothesized role of free radicals and lipid peroxidation in the occurrence of ischaemic brain oedema[10, 11, 33]. As regards the enhanced generation of active oxygens and the occurrence of lipid peroxidation during or following cerebral ischaemia, however, there has been a wide discrepancy between reported experimental results[35]. Since the theory has so far remained controversial, it seems appropriate to examine further whether or not the free radical mechanism plays any significant role in the pathogenesis underlying ischaemic brain oedema.

It has been well substantiated that living cells generate certain amounts of oxygen-free radicals and hydroperoxides[8, 10]. Also, there is no doubt that lipid peroxidation proceeds in incubated brain tissues as evidenced by the increase in TBA-reactive substances[6]. The most essential problem for the "free radical hypothesis" is whether or not an overt peroxidation of membrane lipids is induced by active oxygens at the time of an ischaemic insult or after it, so that the membrane function is irreversibly damaged. Although such a possibility as above has been regarded as unlikely by some authors[18], it still seems possible to analyse the problem from a novel viewpoint which is based on recent knowledge about the metabolism of free arachidonic acid, *i.e.*, the arachidonate cascade.

It is well-known that the brain is rich in the enzymes composing the arachidonate cascade[38]. The production of a variety of biologically potent eicosanoids within the brain was shown to be increased under pathological conditions such as ischaemia. Of particular relevance to the free radical mechanism is the fact that the arachidonate cascade liberates free radicals[27] and that the tissue level of free radicals (ambient free radicals) affects the activity of key enzymes of the cascade such as cyclo-oxygenase and lipoxygenase[28]. Therefore, it can be postulated that an increased tissue level of free radicals may stimulate either one or both of the non-enzymatic (autoxidation) and enzymatic (the arachidonate cascade) pathways of lipid peroxidation. During the past several years, we have concentrated our research effort on the elucidation of the roles of free radicals and the arachidonate cascade in the occurrence of ischaemic brain damage. In the present paper, pertinent findings are summarized to put forward a novel concept on the pathogenesis of ischaemic brain oedema.

The Relevance of the Cyclo-oxygenase Pathway to Brain Oedema and rCBF Following Prolonged Ischaemia and Recirculation

Using the prolonged unilateral and transient bilateral carotid artery occlusion models in gerbils, Gaudet et al.[13, 14] showed a significant elevation of the brain level of prostaglandins (PGs) after ischaemia. The increase in PGs was particularly marked following recirculation, the time course of which was quite parallel to the brain level of free arachidonic acid (AA) as was shown with a similar animal model[39].

Using the middle cerebral artery (MCA) occlusion model in cats, we recently studied the relationship between the regional brain level of each PG and the topographically correlated rCBF values[19]. As shown in Fig. 1, there was a moderate increase in each PG after prolonged ischaemia (PI) only in brain regions where

the mean rCBF during MCA occlusion (clip flow) dropped below 20–25 ml/100 g/min. It has been reported that the rCBF threshold for membrane depolarization, hence intracellular influx of Ca^{++} is less than 10 ml/100 g/min[20]. It is of interest that the rCBF threshold for elevated PG synthesis in the present model was considerably higher than that for membrane depolarization. This is probably because of the inhomogeneity of rCBF reduction within the MCA territory, the activation of phospholipases coupled with receptor stimulation, or diffusion of free arachidonic acid from the focal to the perifocal areas. In any event, the common rCBF threshold for elevated PG synthesis and oedema formation is suggestive of an interrelation between the two phenomena.

The alterations in brain levels of 6-keto-$PGF_{1\alpha}$ and TXB_2 shown in the present study were far less pro-

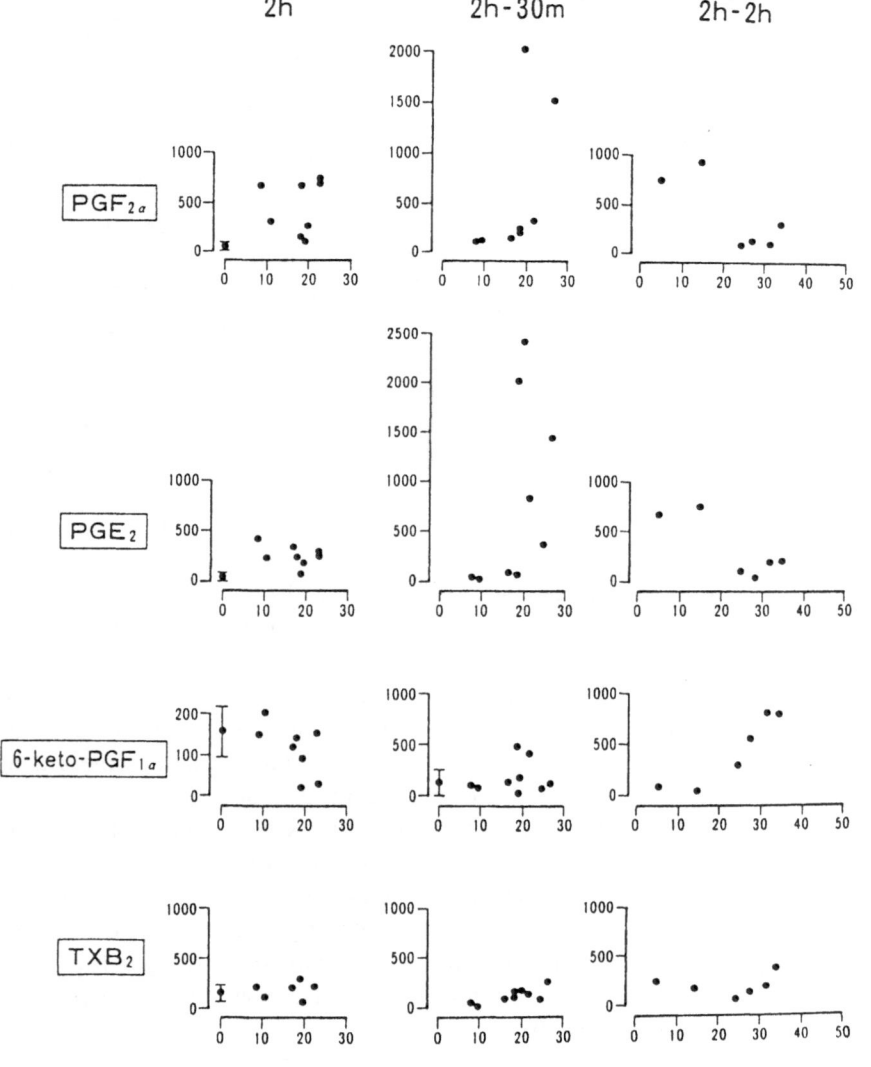

Fig. 1. Prostaglandin content of the cat brain topographically correlated to the mean rCBF during MCA occlusion. Abscissa: prostaglandin level (pg/mg dry weight); ordinate: the mean rCBF during MCA occlusion (ml/100 g/min). *2 h* After 2 hours MCA occlusion; *2 h-30 min* after 30 min recirculation following 2 hours MCA occlusion; *2 h-2 h* after 2 hours recirculation following 2 hours MCA occlusion

nounced than those reported by others. This is probably due to the use in the present study of brain perfusion with cold saline prior to sampling. In contrast to prolonged ischaemia, recirculation after two hour's MCA occlusion was followed by an explosive increase of PGE_2 and $PGF_{2\alpha}$. It should be noted that the elevation of each PG after recirculation was most prominent in the moderately ischaemic areas, *i.e.*, perifocal areas. In the focal areas where the clip flow was less than $15\,ml/100\,g/min$, the PG increase was more delayed and less prominent. In the brain areas where the clip flow was close to zero, no increase in PGs was observed.

Gaudet et al.[14] reported that there was a poor correlation between the brain level of each PG and brain oedema. In the permanent MCA occlusion model in cats used in the present study, the rCBF threshold for oedema formation was found to be 20–$25\,ml/100\,g/min$[22]. Below this rCBF threshold, there was an inverse relationship between the clip flow and the magnitude of cortical oedema. In prolonged ischaemia, no correlation was found between the level of each PG and the clip flow. Therefore, like the study of Gaudet et al., the present study does not support the view that the enhanced PG-synthesis is directly related to oedema formation. However, it cannot be ruled out that the pronounced rise in PGE_2 and $PGF_{2\alpha}$ observed in the perifocal area after recirculation may be relevant to aggravation of oedema in the same area.

We next examined the effect of a cyclo-oxygenase inhibitor, indomethacin ($4\,mg/kg$) on ischaemic brain oedema (the cortical specific gravity) and rCBF using the cat MCA occlusion model[3, 22, 29]. In prolonged ischaemia, brain oedema was rather worsened by administration of indomethacin (Fig. 2) whereas the rCBF during MCA occlusion was not affected (Fig. 3). This result is in agreement with preceding studies showing that the cyclo-oxygenase products do not play a major role in oedema formation due to prolonged ischaemia. On the other hand, indomethacin markedly reduced brain oedema, at the same time improving rCBF after recirculation (Figs. 2 and 3). Parallel studies revealed that free radical scavengers such as ONO 3141 and AVS had effects similar to that of indomethacin with respect to brain oedema and rCBF[2, 22]. Since these free radical scavengers do not prevent PG synthesis, the deleterious effect of recirculation on brain oedema may be ascribed not to enhanced synthesis of PGs but to generation of free radicals or oxidants.

With respect to postischaemic hypoperfusion, observed alterations in the brain levels of 6-keto-$PGF_{1\alpha}$ and TXB_2 do not provide any clue as to their possible involvement. Since there was no significant increase in the brain level of TXB_2 throughout of the experiment, the involvement TXA_2 in the occurrence of postischaemic hypoperfusion appears unlikely. Further, the administration of an inhibitor of TXA_2 synthase, OKY 1581 in the present model showed no significant effect on postischaemic rCBF (unpublished data). By the same token, there was a poor relationship between rCBF and the brain of 6-keto-$PGF_{1\alpha}$. Since the administration of indomethacin, which inhibits the synthesis of PGI_2, significantly improved postischaemic rCBF, it seems unlikely that the diminished synthesis of PGI_2 is in any way involved in the occurrence of postischaemic hypoperfusion. In addition, the brain levels of vasoconstrictive PGs, *i.e.*, $PGF_{2\alpha}$ and PGE_2 showed a reverse relationship to the

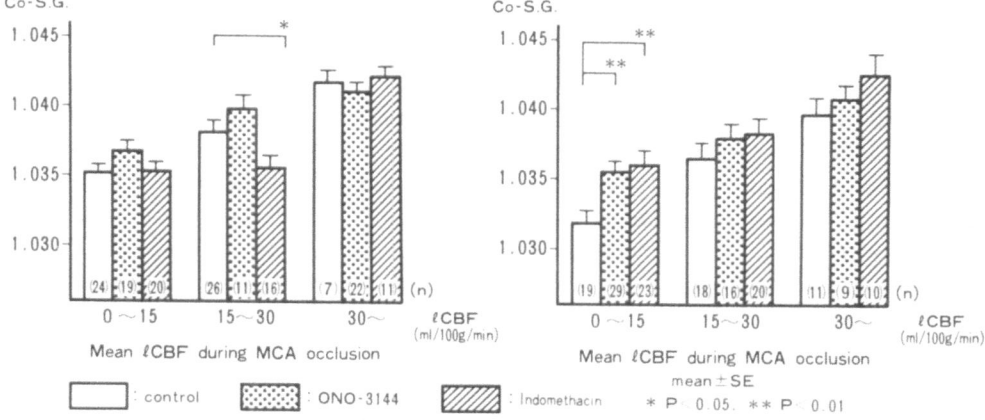

Fig. 2. The cortical specific gravity (Co-SG) topographically correlated to the mean rCBF during MCA occlusion in cats. Data were assembled according to the ranges of the mean rCBF during MCA occlusion, *i.e.*, 0–15 (focal area), 15–30 (perifocal area), and >30 (normal area) ml/100 g/min. Results in the prolonged ischemia (4 hours continuous MCA occlusion) and in the recirculation (2 hours MCA occlusion plus 2 hours recirculation) are shown in the right and left figures, respectively

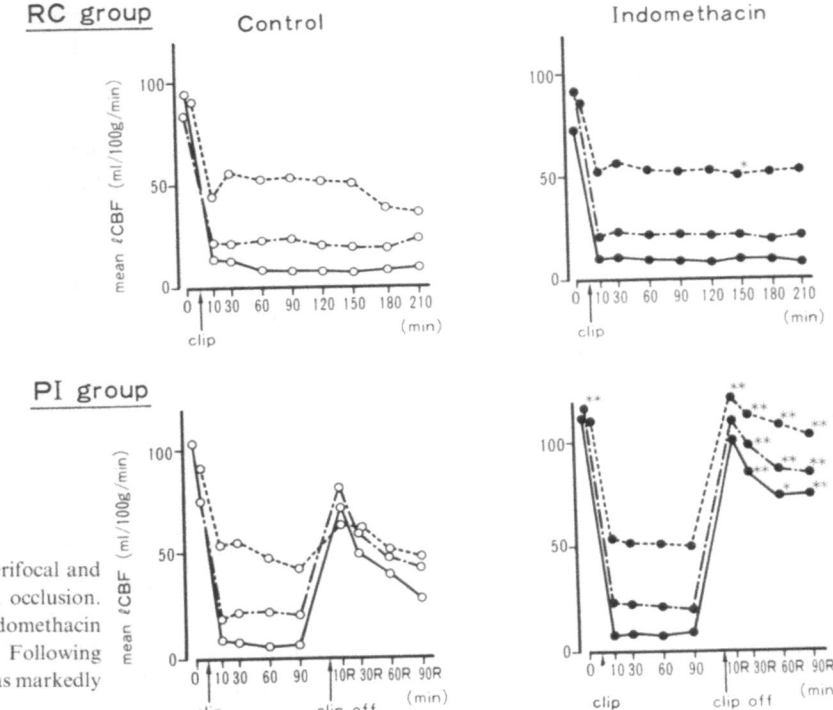

Fig. 3. The time course of rCBF in the focal, perifocal and normal areas in the cat brain exposed to MCA occlusion. During MCA occlusion for 4 hours (PI group), indomethacin (4 mg/kg) caused no alteration in the rCBF. Following recirculation, the postischaemic hypoperfusion was markedly improved by indomethacin

postischaemic course of rCBF. Thus it may be stated that the altered synthesis of each PG has little relevance to the occurrence of postischaemic hypoperfusion. In this regard, attention needs to be directed to the pathogenetic role of oxidants released from the arachidonate cascade which in some unknown way produce damage to the cerebral vasculature and parenchymal cells upon recirculation.

The Relevance of Lipoxygenase Products

It has been shown that the intracerebral injection of AA causes brain oedema[1, 9, 37], which is not suppressed by administration of indomethacin[9]. Likewise, brain oedema following prolonged ischaemia is not ameliorated by indomethacin although accumulation of free arachidonic acid within the brain is evident. Therefore, it seems rather unlikely that metabolites of arachidonic acid via the cyclo-oxygenase pathway are involved in the occurrence of oedema following prolonged ischaemia. Free fatty acids including arachidonic acid may interfere with the energy production in mitochondria and thus may be related to brain oedema[32]. In the face of recent knowledge on the potent phlogistic properties of lipoxygenase products of arachidonic acid[31], however, it is tempting to speculate that the

lipoxygenase products might be involved in the occurrence of brain oedema after prolonged ischaemia.

In this respect, there is as yet very little in the literature concerning the brain level and the physiological or pathological role of each lipoxygenase product in the normal or ischaemic brain. Although a transient increase in leukotrienes (LTs) C_4 and D_4 was reported in the brains of animals exposed to transient ischaemia[30], SAH, or trauma[24], LTs represent only a small portion of the whole spectrum of lipoxygenase products. Therefore, we undertook a study to determine the brain level of each lipoxygenase product (hydroxyeicosatetraenoic acid: HETE) using the rat MCA occlusion model[36].

The extent of cerebral infarction and the time-course of hemispheric oedema following unilateral permanent MCA occlusion in rats were previously described[16]. The content of each HETE in the affected hemisphere was determined by the use of the high performance liquid chromatogrphy (HPLC). As shown in Table 1, a significant and general increase of HETEs was revealed as late as 72 hours after MCA occlusion. Although hemispheric oedema became maximal at the same period in this model, it is not clear from this result whether or not the increased synthesis of lipoxygenase products preceded or paralleled oedema formation. It would even be

Table 1. *The Hemispheric Content of HETEs Before and After MCA Occiusion in Rats (ng/g Wet Weight)*

Duration of ischaemia		5-HETE	9-HETE	8- and/or 12-HETE	11-HETE	15-HETE
0 (control)	(n = 4)	0.0 ± 0.0	0.0 ± 0.0	0.0 ± 0.0	0.0 ± 0.0	2.1 ± 2.1
24 hours	(n = 4)	0.0 ± 0.0	0.0 ± 0.0	0.0 ± 0.0	1.6 ± 1.6	9.1 ± 3.8
72 hours	(n = 4)	18.3 ± 2.8**	16.6 ± 3.0*	23.7 ± 14.9	14.0 ± 3.4*	19.8 ± 5.5*

Each value represents the mean ± SE for the number of independent determination indicated in the parentheses. * $p < 0.05$ and ** $p < 0.01$. Significantly different from the control.

possible to interprets the data as indicating the occurrence of lipid peroxidation in the infarcted brain. Therefore, we further proceeded to study the alterations in the synthesis of lipoxygenase products after MCA occlusion, within a particular fraction of the brain, *i.e.*, brain microvessels.

The Synthesis of Lipoxygenase Products in Brain Microvessels

Since the major function of the blood-brain barrier (BBB) resides in endothelial cells, it goes without saying that brain microvessels (MV) are involved in the pathogenetic mechanism underlying ischaemic brain oedema. The metabolically active MV was prepared from affected hemispheres by the use of nylon meshes and sucrose-density centrifugation. The eicosanoid synthetic capacity of the MV, *i.e.*, the conversion ratio from radiolabelled AA to each eicosanoid, was determined by radiochromatography. A generalized increase in the synthetic capacity was revealed with each eicosanoid and it was most pronounced with hydroxy-acids (HETEs)[4]. Of importance is the fact that the synthesis of HETEs was greatest at 24 hours after MCA occlusion, at which time the hemispheric oedema had been found to develop at a maximal speed[17] (Fig. 4). Such a result as above would allow an interpretation that lipoxygenases within a particular fraction of the brain such as the MV, are stimulated early in the course of cerebral ischaemia and participate in oedema formation.

Regarding the chemical mechanisms involved in the stimulation of lipoxygenase activity, we have shown that a hydroperoxyeicosatetraenoic acid, 15-HPETE potently stimulates lipoxygenases of the brain MV[25]. Other compounds hitherto known as oedema factors such as monoamines showed no significant effect in this respect. A free radical scavenger, AVS [1,2-bis(nicotinamide)-propane, Chugai Pharamceut. Co.] inhibited the 15-HPETE-induced production of lipoxygenase products of the MV *in vitro*[25], and markedly mitigated brain oedema following MCA occlusion in cats[2] as well as in rats[16]. From these pieces of evidence, it may be conjectured that cerebral ischaemia first leads to an increase in the ambient level of hydroperoxides and/or oxygen-free radicals, which in turn cause the activation of a lipoxygenase pathway within the brain MV. Such a speculation as above led us to the next question: in what way is the activation of lipoxygenase(s) of the brain MV linked to formation of the brain oedema?

The Effect of 15-HPETE on the Membrane-bound Enzyme, Na^+,K^+-ATPase

It has been widely accepted that lipid peroxidation is harmful to cells because it damages the membrane-bound enzymes, such as Na^+,K^+-ATPase[10, 12]. Insomuch as the arachidonate cascade is a form of lipid peroxidation, our study showing that the lipoxygenase products participate in oedema formation is apparently consonant to the above thesis. But, the study which was carried out to examine the possible damaging effect of 15-HPETE on Na^+,K^+-ATPase of the MV and synaptosomes revealed quite an unexpected result[26]. In relatively low concentrations ($< 10^{-5}$ M), 15-HPETE was found to increase the activity of the enzyme, whereas the agent significantly suppressed the enzyme activity of synaptosomes in higher concentrations. This enhancing effect of 15-HPETE on MV Na^+,K^+-ATPase was considered to be due to stimulation of the phospholipase A_2 and the lipoxygenase pathway because its effect was completely inhibited by quinacrine, ETYA, or caffeic acid[26] (Fig. 5). Indomethacin did not inhibit the effect of 15-HPETE, implicating that the cyclo-oxygenase pathway is not involved in the activation of MV Na^+,K^+-ATPase. In any event, results of this study indicate that the activity of MV Na^+,K^+-ATPase is more likely to be enhanced than suppressed in the case of an elevation of hydroperoxides due to any pathological conditions, since any hydroperoxide would hardly exceed the concentration level of 10^{-5} M in the living brain.

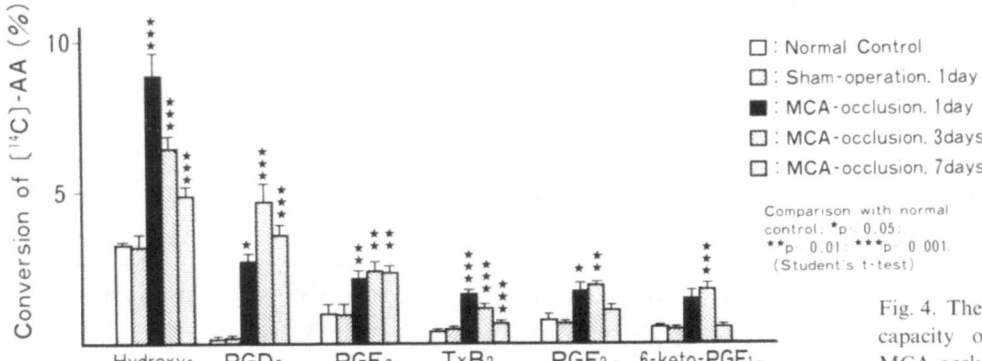

□ : Normal Control
□ : Sham-operation, 1 day
■ : MCA-occlusion, 1 day
□ : MCA-occlusion, 3 days
□ : MCA-occlusion, 7 days

Comparison with normal
control. *p < 0.05;
**p < 0.01; ***p < 0.001.
(Student's t-test)

Fig. 4. The alteration of eicosanoid synthetic capacity of brain microvessels following MCA occlusion

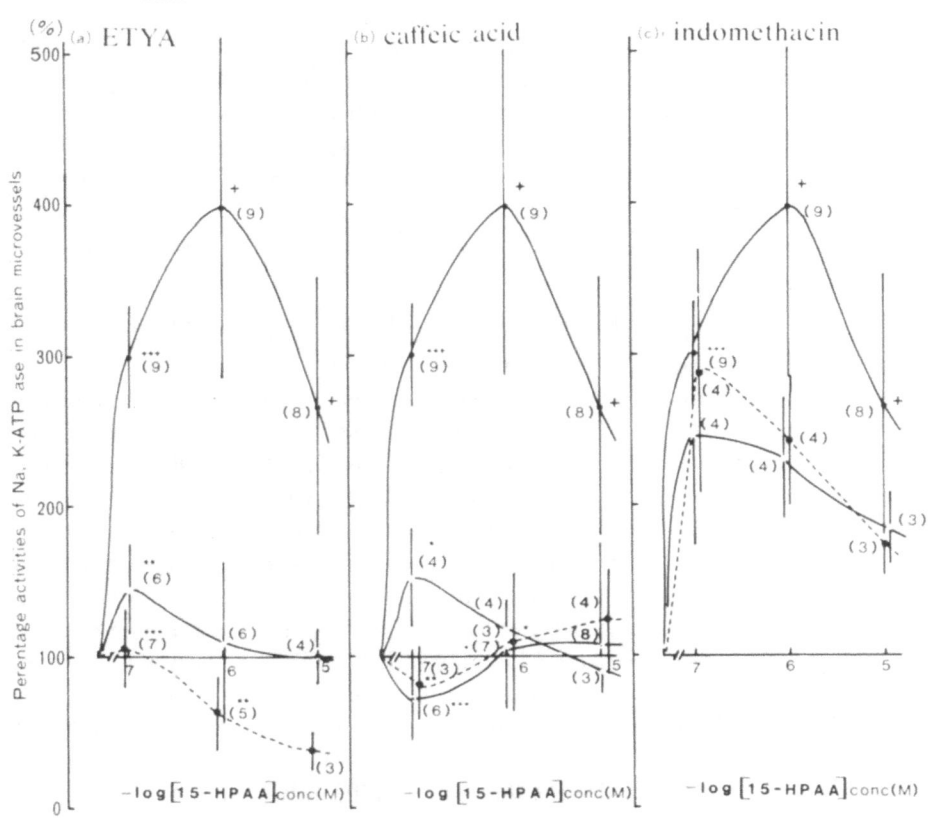

Fig. 5. The effects of ETYA, caffeic acid, and indomethacin on the 15-HPETE (HPAA)-induced enhancement of the Na^+,K^+-ATPase activity of rat brain microvessels. The top curve in each figure shows the dose-dependent enhancement of Na^+,K^+-ATPase activity by 15-HPETE. Numbers in parentheses indicate the number of experiments

The Behaviour of BBB in the Ischaemic Brain

The above findings led us to re-evaluate the role of sodium entry across the BBB in the pathogenesis of brain oedema. As already reported, a firm coupling between the sodium entry and the water entry was revealed throughout the evolution of ischaemic brain edema following permanent MCA occlusion in rats[17] (Fig. 6). Almost a similar linear correlation between the brain contents of sodium and water has been found also with the cat cerebral cortex subjected to MCA occlusion[21, 33].

Such a result as above is predictable on the basis of Gibbs-Donnan equilibrium between the blood and the

cerebral extracellular fluid (ECF). If the BBB behaves like a semipermeable membrane after the onset of ischaemia, the sodium concentration of oedema fluid is expected to be slightly lower than that in serum. Assuming that the sodium concentration in the oedema fluid is 140 mEq/L and that the one-to-one exchange of sodium with potassium is not accompanied by water movement, the following equation exists[23]:

$$dH_2O = \frac{dNa - dK}{140} \; (1/kg \; dry \; weight)$$

where dH_2O, dNa, dK represent the increments or decrements of hemispheric contents of H_2O, sodium, and potassium.

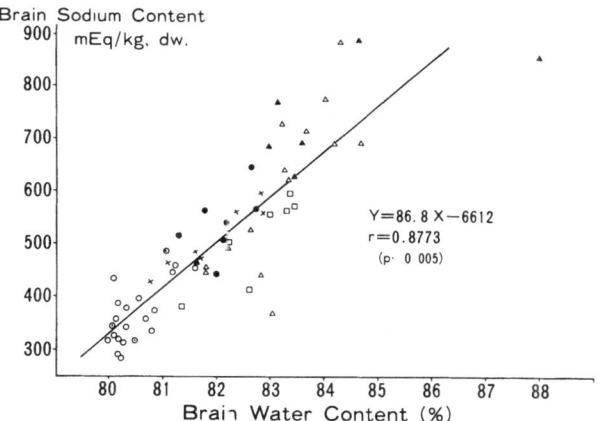

○ : 3 - 6 h ; ⊙ : 12 h ; ● : 1 day ; △ : 2 day ; ▲ : 3 day ; □ : 4 day ; X : 7 day

Fig. 6. The correlation between the brain sodium content and the brain water content in rat hemispheres exposed to permanent MCA occlusion

According to this equation, the theoretical value of dNa (dNa$_t$) can be calculated with each specimen (rat hemispheres subjected to permanent MCA occlusion) using the actual values of dH$_2$O and dK. The relationship between dNa$_t$ and the actual value of dNa was examined, revealing a highly significant linear correlation as shown in Fig. 7. Such a correlation provides

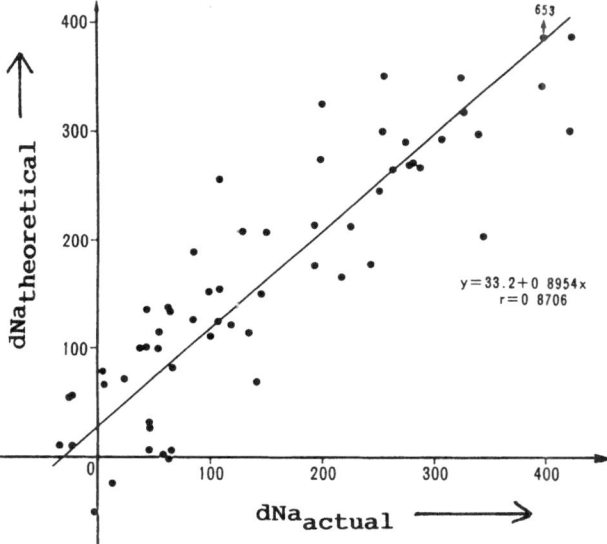

Fig. 7. The correlation between the theoretical and actual values of dNa

strong support to the classical concept that the BBB behaves as if it were a semipermeable membrane following ischaemia and the oedema fluid can be regarded as the ultrafiltrate of serum in accordance to Gibbs-Donnan equilibrium. Nevertheless, would it necessarily follow that the BBB, *i.e.*, endothelial cells

become freely permeable to sodium and other ions as soon as ischaemia starts?

The possible route(s) of sodium entry across the BBB would be through the opened tight junction or through the endothelial cells. It seems quite plausible that the free passage of sodium occurs across the BBB in the presence of frank destruction of endothelial cells and tight junctions as observed in the infarcted brain. However, in the relatively early phase of ischaemia, especially in the perifocal areas, tight junctions remain closed at least to proteins and there is little morphological evidence for destruction of endothelial cells[23]. Brain oedema developing under these conditions has been designated as cytotoxic oedema. It must be stressed that even during the stage of cytotoxic oedema, there is a significant increase in the brain contents of sodium so far as the ischaemic brain oedema is concerned. Therefore, the sodium entry during the early phase of ischaemia may not occur by means of simple diffusion through the disrupted BBB as it occurs during the later stage of vasogenic oedema. Involvement of alternative pathways, such as the sodium transport mechanism(s) in endothelial cells should then be considered.

In this regard, it has recently been suggested that Na$^+$,K$^+$-ATPase is located only in the abluminal membrane of endothelial cells and it functions to transport sodium in a unipolar direction[7, 15]. This intriguing idea may be coupled with our own finding to form a hypothesis that the activity of endothelial Na$^+$,K$^+$-ATPase is enhanced by an elevated level of hydroperoxides following ischaemia, which results in an increased transport of sodium and water across the BBB, hence the formation of brain oedema.

To verify this novel concept, the following experiment was carried out. Using cats, the MCA was transorbitally clipped at its origin and the distal part of the MCA was cannulated. Through the cannula, the MCA territory was periodically perfused with aerated Krebs-Ringer buffer solution containing 10^{-5}M ouabain (Fig. 8). In this experiment, we aimed to inhibit endothelial Na$^+$,K$^+$-ATPase within the ischaemic MCA territory with a minimal amount of ouabain systemically distributed. As shown in Fig. 9, oedema in the MCA territory was significantly lessened by the intra-arterial infusion of ouabain[5]. Further, the BUI of ^{22}Na was measured according to the Oldendorf's method. In the control group, the BUI of sodium was increased as the ischaemic flow was reduced below 30 ml/100 g/min. This increase in BUI was clearly inhibited by administration of ouabain[33]. It seems clear from these results that the activity of endothelial

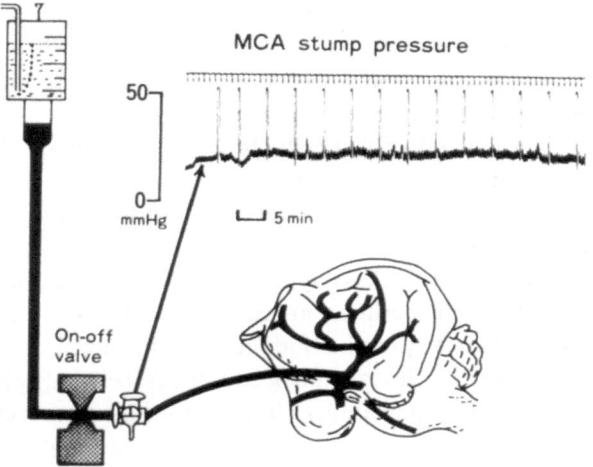

Fig. 8. The experimental design of the intermittent, intra-arterial perfusion of the MCA territory

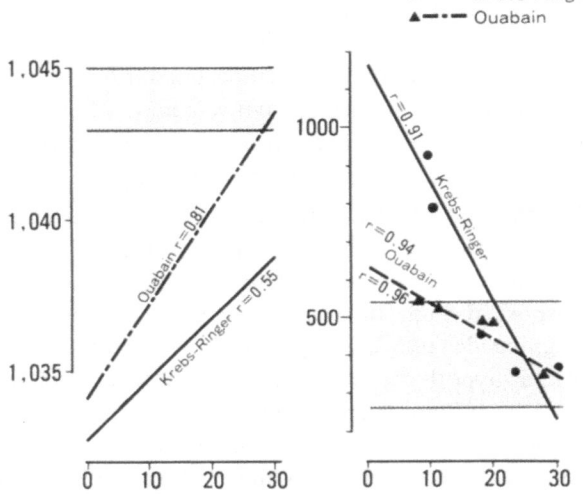

Fig. 9. Effects of ouabain on the cortical specific gravity and the cortical sodium content following intermittent, intra-arterial perfusion in the cat MCA occlusion model. Lines designate the linear correlation between the mean rCBF following MCA occlusion and the topographically corresponding cortical specific gravity in each experimental group

Na^+,K^+-ATPase in the ischaemic MCA territory is causally related to sodium influx as well as oedema formation in the same brain area. We also showed that AVS, which inhibits the 15-HPAA-induced enhancement of Na^+,K^+-ATPase *in vitro*[26], significantly suppressed oedema formation in the same experimental model[33].

A Novel Concept on the Pathogenetic Mechanism Underlying Ischaemic Brain Oedema

Hitherto obtained experimental results are in support of the hypothesis that following regional, incomplete ischaemia, the lipoxygenase activity within the brain MV is enhanced presumably because of an increase in the ambient level of free radicals and hydroperoxides. Subsequently, the activity of endothelial Na^+,K^+-ATPase is enhanced, leading to an increased transport of sodium together with water from blood to brain (Fig. 10). This mechanism would be in operation in brain areas where endothelial cells remain viable in the face of reduced blood supply and can supply sufficient ATP for its Na^+,K^+-ATPase to function.

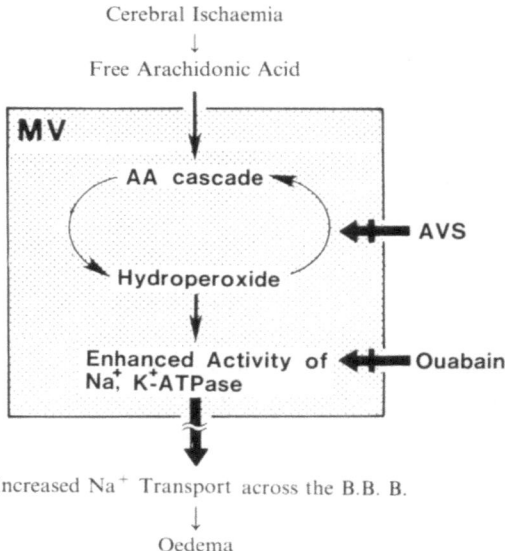

Fig. 10. The proposed biochemical mechanism underlying ischemic brain edema. The anti-oedema effects of AVS and ouabain are attributable to their suppressive effects on the formation of hydroperoxide and the activity of Na^+,K^+-ATPase within brain microvessels (*MV*), respectively

It must be stressed, however, that the event elicited by recirculation is slightly different. The aggravation of brain oedema and the postischaemic hypoperfusion following recirculation appears to be due to the prominent activation of the cyclo-oxygenase pathway. The oxidant released from PGG_2 may be the factor responsible for the detrimental effects of recirculation, which causes more or less non-specific damage of endothelial and parenchymal cells.

From the above point of view, brain oedema in either permanent ischaemia or recirculation can no longer be regarded as "an inevitable accompaniment of necrosis". Rather, ischaemic brain oedema may be at least partially mitigated by the use of appropriate drugs which eliminate relevant chemical factors. As exemplified by the effects of indomethacin and free radical scavengers such as AVS, further investigation on the pharmacological control of ischaemic brain oedema seems to be warranted.

References

1. Aritake K, Wakai S, Asano T, Takakura K (1983) Peroxidation of arachidonic acid and brain edema. Brain Nerve 35: 965–973

2. Asano T, Johshita H Koide T, Takakura K (1984) Amelioration of ischaemic cerebral oedema by a free radical scavenger, AVS, 1,2-Bis(nicotinamide)-propane. An experimental study using a regional ischaemia model in cats. Neurol Res 6: 163–168

3. Asano T, Matsui T, Basugi N, Tamura A, Takakura K, Sano K (1984) The effect of indomethacin on cortical specific gravity during regional ischemia and recirculation. In: Go KG, Baethmann A (eds) Recent progress in the study and therapy of brain edema. Plenum Press, New York, pp 617–626

4. Asano T, Gotoh O, Koide T, Takakura K (1985) Ischemic brain edema following occlusion of the middle cerebral artery in the rat. II. Alteration of the eicosanoid synthesis profile of brain microvessels. Stroke 16: 110–113

5. Asano T, Shigeno T, Hanamura T, Koide T, Matsushita H, Watanabe E, Mima T, Johshita H, Usui M, Takakura K (1985) Alteration of brain capillary function in cerebral ischemia: Role of capillary Na^+, K^+-ATPase in ischemic edema formation. J Cereb Blood Flow Metabol 5 [Suppl 1]: S 63–S 64

6. Barber AA, Bernheim F (1967) Lipid peroxidation: its measurement, occurrence, and significance in animal tissues. Adv Gerontol Res 2: 355–403

7. Betz AL, Firth JA, Goldstein GW (1980) Polarity of the blood-brain barrier: Distribution of enzymes between the luminal and antiluminal membranes of brain capillary endothelial cells. Brain Res 192: 17–28

8. Boveris A (1977) Mitochondrial production of superoxide radical and hydrogen peroxide. Adv Exp Biol Med 78: 67–82

9. Chan PH, Fishman RA, Caronna J, Schmidley JW, Prioleau G, Lee J (1983) Induction of brain edema following intracerebral injection of arachidonic acid. Ann Neurol 13: 625–632

10. Del Maestro RF (1980) An approach to free radicals in medicine and biology. Acta Physiol Scand [Suppl] 492: 153–169

11. Demopoulos HB, Flamm ES, Seligman ML, Poser R, Pietronigro O, Ransohoff J (1975) Molecular pathology of lipids in CNS membranes. In: Joebsis FF (ed) Oxygen and physiological function. Professional Information Library, Dallas, pp 491–508

12. Demopoulos HB, Flamm ES, Seligman ML, Mitamura JA, Ransohoff J (1979) Membrane perturbations in CNS injury: theoretical basis for free radical damage and a review of the experimental data. In: Popp AJ, Bourke RS, Nelson LR, Kimelberg HK (eds) Neural trauma. Raven Press, New York, pp 63–78

13. Gaudet RJ, Alam I, Levine L (1980) Accumulation of cyclo-oxygenase products of arachidonic acid metabolism in gerbil brain during reperfusion after bilateral common carotid artery occlusion. J Neurochem 35: 653–658

14. Gaudet RJ, Levine L (1980) Effect of unilateral common carotid artery occlusion on levels of prostaglandins D_2, $F_{2\alpha}$, and 5-keto-prostaglandin $F_{1\alpha}$ in gerbil brain. Stroke 11: 648–652

15. Goldstein GW, Betz AL (1983) Recent advances in understanding brain capillary function. Ann Neurol 14: 389–395

16. Gotoh O, Koide T, Asano T, Takakura K, Tamura A, Sano K (1984) A model to study ischemic brain edema in rats and the influence of drugs. In: Go KG, Baethmann A (eds) Recent progress in the study and therapy of brain edema. Plenum Press, New York, pp 499–508

17. Gotoh O, Asano T, Koide T, Takakura K (1985) Ischemic brain edema following occlusion of the middle cerebral artery in the rat. I. The time courses of brain water, sodium, and potassium contents and blood-brain barrier permeability to ^{125}I-albumin. Stroke 16: 101–109

18. Halliwell B, Gutteridge JMC (1984) Lipid peroxidation, oxygen radicals, cell damage, and antioxidant therapy. Lancet 23: 1396–1397

19. Hanamura T, Asano T, Shigeno T, Mima T, Takakura K (1986) Prostaglandin profiles in relation to local circulatory changes following focal cerebral ischemia in cats. Paper presented in 6th International Conference on Prostaglandins and Related Compounds, in Florence, Italy, 1986. Paper in preparation

20. Harris RJ, Symon L, Branston NM, Bayhan M (1981) Changes in extracellular calcium activity in cerebral ischaemia. J Cereb Blood Flow Metabol 1: 203–209

21. Hossmann KA (1985) The pathophysiology of ischemic brain edema. In: Inaba Y, Klatzo I, Spatz M (eds) Brain edema. Springer, Berlin Heidelberg New York Tokyo, pp 367–384

22. Johshita H, Asano T, Takakura K, Experimental evaluation of the role of cyclo-oxygenase pathway in the pathogenesis of ischaemic brain edema. Paper in preparation

23. Katzman R, Clasen R, Klatzo I, Meyer JS, Pappius HM, Waltz AG (1977) Report of joint committee for stroke resources. IV. Brain edema in stroke. Study group on brain edema in stroke. Stroke 8: 512–540

24. Kiwak KJ, Moskowitz MA, Levine L (1985) Leukotriene production in gerbil brain after ischemic insult, subarachnoid haemorrhage, and concussive injury. J Neurosurg 62: 865–869

25. Koide T, Gotoh O, Asano T, Takakura K (1985) Alterations of the eicosanoid synthethic capacity of rat brain microvessels following ischemia: Relevance to ischemic brain edema. J Neurochem 44: 85–93

26. Koide T, Asano T, Matsushita H, Takakura K (1986) Enhancement of ATPase activity by a lipid peroxide of arachidonic acid in rat brain microvessels. J Neurochem 46: 235–242

27. Kuehl FA, Egan RW (1980) Prostaglandins, arachidonic acid, and inflammation. Science 210: 978–984

28. Lands WEM, Kulmacz RJ, Marshall PJ (1984) Lipid peroxide actions in the regulation of prostaglandin biosynthesis. In: Pryor WA (ed) Free radicals in biology, vol 6. Academic Press, Orlando, pp 39–63

29. Matsui T, Basugi N, Asano T, Takakura (1984) The effect of indomethacin on ischemic brain edema: a study using cat middle cerebral artery occlusion combined with recirculation. Neurol Med Chir 24: 5–12

30. Moskowitz MA, Kiwak KJ, Hekiman K, Levine (1984) Synthesis of compounds with properties of leukotrienes C 4 and D 4 in Gerbil brains after ischemia and reperfusion. Science 224: 886–889

31. Samuelsson B (1983) Leukotrienes: mediators of immediate hypersensitivity reactions and inflammation. Science 220: 568–575

32. Sato K, Yamaguchi M, Mullan S, Evans JP, Ishii S (1969) Brain edema. A study of biochemical and structural alteration. Arch Neurol 21: 413–424

33. Shigeno T, Asano T, Hanamura T, Watanabe E, Mima T, Usui M, Tagusagawa Y, Ochiai C, Takakura K (1985) Challenge to brain capillary function in focal cerebral ischemia. Presented in Symposium on the Blood-Brain Barrier. Official Satellite Symposium to the 12th International Symposium on Cerebral Blood Flow and Metabolism, Copenhagen, 1985. Paper in submission

34. Siesjo BK (1981) Cell damage in the brain: a speculative synthesis. J Cereb Blood Flow Metabol 1: 155–185

35. Siesjo BK, Wieloch T (1985) Brain injury: Neurochemical aspects. In: Becker DP, Povlishock JT (eds) Central nervous system trauma status report. William Byrd Press, Richmond, VA, pp 513–532

36. Usui M, Asano T, Takakura K (1987) Identification and quantitative analysis of hydroxyeicosatetraenoic acids in the rat brain exposed to regional ischemia. Stroke 18: 490–494

37. Wakai S, Aritake K, Asano T, Takakura K (1982) Selective destruction of the outer leaflet of the capillary endothelial membrane after intracerebral injection of arachidonic acid. Acta Neuropathol 58: 303–306

38. Wolfe LS (1982) Eicosanoids: prostaglandins, thromboxanes, leukotrienes, and other derivatives of carbon-20 unsaturated fatty acids. J Neurochem 38: 1–14

39. Yoshida S, Inoh S, Asano T, Sano K, Kubota M, Shimaziki H, Ueta N (1980) Effect of transient ischemia on free fatty acids and phospholipids in the gerbil brain; lipid peroxidation as possible cause of postischaemic injury. J Neurosurg 53: 323–331

Correspondence: Prof. T. Asano, M.D., D.M.Sc., Department of Neurosurgery, Saitama Medical Center, Saitama Medical School, 1981 Tsujido, Kamota, Kawagoe, Saitama 350, Japan.

Acta Neurochirurgica, Suppl. 41, 95–96 (1987)

Discussion

S. Ishii

Department of Neurosurgery, Juntendo University, Tokyo, Japan

I will talk today about the ill effects of oedema on the basis of our own experimental studies.

It is quite clear that extensive cerebral oedema produces a rise in ICP and causes cerebral herniation. These are some of the obvious ill effects of cerebral oedema, but today I will discuss the direct effects of brain oedema upon the brain tissue itself.

First, I will deal with hypertension and oedema. Hypertension was induced in cats by inflating a balloon situated in the aorta. Sustained hypertension ranging from 200–250 mm Hg for up to thirty minutes was easily obtained without any significant complication in the animal, other than in the brain. After hypertension lasting 30 minutes, multifocal lesions were found which had been stained with Evans blue at the watershed areas between major cerebral arteries.

The lesions were usually confined to the superficial cortical grey, and dye very rarely extended further into the white matter. It was also confirmed that leakage of macromolecules through the BBB ceased within twelve hours. These lesions may be called cortical oedema.

When the animals were subjected to both hypertension and craniectomy, these changes became intensified and the Evans blue which stained cortical lesions migrated further into the white matter.

When animals subjected to both hypertension and craniectomy were allowed to survive for 72 hours, extensive brain oedema, associated with the shift of the midline structures, was observed on the side of the craniectomy.

It has been widely accepted that exchange of water and ions between blood and brain is physiologically controlled by the following three factors,

1) the hydrostatic pressure gradient between blood and brain parenchyma,

2) the osmotic pressure gradient between these two, and

3) energy dependent active transport of water and ions by the endothelial and glial cells.

In the afore-mentioned experimental models, leakage of plasma contents occurs from a structurally normal vessel wall into metabolically normal brain tissue at least in its initial stage. Therefore it is reasonable to assume that the changes of the hydrostatic pressure which have developed between blood and brain parenchyma may be playing a role in the leakage of macromolecules. This type of oedema may be called "hydrostatic edema".

In the hypertension model luminal pressure on both the arterial and venous side may increase. In the craniectomy-hypertension model, elevation of luminal pressure and decrease in tissue pressure may occur at the same time.

Why then, does focal oedema occur at the watershed areas? Extravasation of dye into the cortex in the watershed area could be seen through the dura immediately after the start of hypertension in the craniectomy-hypertension model.

The mechanism of focal oedema may be as follows.

With the rapid elevation of perfusion pressure following hypertension, transient vascular engorgement may occur. The tissue pressure at an area supplied by the major arteries may then rise higher than in the relatively avascular watershed area.

The pressure gradient between blood and brain tissue becomes greater in the avascular watershed area. Thus the leakage of plasma contents may be more likely to occur. The next problem is why the extensive hemispheric oedema associated with midline shift develops in the later hours after hypertension and craniectomy. As was mentioned earlier, focal brain oedema produced inside of the closed skull by simple hypertension usually subsides within 12 hours.

In the presence of a bone defect the mechanism of

rapid tissue pressure equilibrium existing in brain when it is encased entirely within the skull, no longer exists. The tissue pressure of the brain on the side of the craniectomy may remain lower compared to that of the other side, and more extensive leakage of macro-molecules may occur on the craniectomized side. In fact, under conditions such as this, electron micro-scopic study demonstrated that the extracellular space had been widely distended and filled with the reaction mixture of HRP. If the amount of extravasated oedema fluid is large enough, it will persist for a long time because of the water-retaining ability to the serum proteins.

This long-lasting abnormal extracellular environment will result in the secondary metabolic impairment of neural tissue and the cytotoxic type of brain oedema may then develop.

In any event it is important to stress that oedema *per se* can cause the development of secondary oedema.

The unfavorable effects of this initial oedema are also very frequently encountered in clinical neurosurgery. In this context the results of experiments using the model of intracerebral haematoma proved to be of interest. As a model of intracerebral haematoma a small amount of blood was injected into the brain through a burr-hole.

After this injection of blood, if the animals were subjected to induced hypertension for 30 minutes, a slight leakage of dye around the haematoma was found. When cats underwent haematoma removal through a burr-hole some time after the injection of blood and during the period of hypertension, a moderate degree of dye was observed after 48 hours. If cats were subjected to hypertension, craniectomy and haematoma removal after the blood injection, the changes became more obvious. Marked oedema with a shift of the midline can be observed. Here again, you may see that focal brain oedema initiated mainly by hydrostatic factors, such as

haematoma removal, hypertension and craniectomy, may result in extensive oedema.

A similar phenomenon can be seen in cerebral infarction. The middle cerebral artery of a cat was clipped when in either a normotensive or hypertensive state. The differences between the two were quite obvious. In the hypertensive cat, leakage of the vital dye occurred more extensively and it extended beyond the border of infarction. When these brains are examined histologically, it is seen that the initial leakage of oedema fluid occurred in an area beyond the infarcted zone caused progressive impairment of microcir-culation. After this additional oedema develops, *i.e.* a peripheral extension of the oedema.

The oedema associated with congenital hydroceph-alus will be very briefly mentioned. An HTX albino rat was kindly given by Professor Kohn of the University of Texas. When low molecular weight tracers such as dextran or HRP were introduced into the lateral ventricle of the hydrocephalic rat, because of the aqueduct stenosis the dye penetrated into the brain parenchyma. It then drained into either the subarach-noid space or lymphatic channels of epipharynx or nasal mucosa.

In the rat the white matter already looked oedema-tous after one day. After seven days, liquefaction of the white matter started and in 21 days an intracerebral cyst developed and the white matter had virtually disap-peared. It has been believed that relatively mild mor-phological changes are noticeable in the cortex even in extensive hydrocephalus. However, if we carefully examine by EM or special staining techniques, obvious changes in the neurocytes, dendrites, synapses or myelin sheath may be found. These are the ill effects of oedema produced by CSF.

Correspondence: Dr. Shozo Ishii, Department of Neurosurgery, Juntendo University of Medicine, 2-1-1- Hongo, Bunkyo-ku, Tokyo 113, Japan.

Acta Neurochirurgica, Suppl. 41, 97–103 (1987)
© by Springer-Verlag 1987

Recovery of Brain Function After Ischaemia

L. Symon

Gough-Cooper Department of Neurological Surgery, Institute of Neurology, London, U.K.

Summary

Experimental evidence has recently suggested that early reperfusion following at least focal cerebral ischaemia is accompanied by a return of function which has apparently been suspended during the ischaemic period. The experimental evidence for this is presented.

Clinical correlates of this reversible ischaemia sometimes referred to as "penumbral ischaemia" are well known in relation to aneurysm surgery. Several examples are presented in this paper. It is also clear that less easily documented and verifiable recovery from long-term ischaemia may occur in neurosurgery and in interesting case suggestive of this is presented. It involved a middle cerebral occlusion which occurred during the excision of a large meningioma.

Keywords: Ischaemic thresholds; penumbra; reperfusion functional recovery.

Introduction

Complete and permanent arrest of the cerebral circulation leads to death. Within seconds of such complete arrest there is depression of the brain electrical activity and within a few minutes gross disruption of the normal energy metabolism within the brain has occurred with failure of ionic homeostatic mechanisms. Such disruption produces irreversible cell change and death follows within 5–10 minutes[9, 12, 34, 40]. In recent years, however, considerable interest has developed in the potential viability of neuronal function after very much more protracted periods of ischaemia than have been thought possible[17, 28]. In clinical neurosurgery it has also been clear that patients with established cerebral vascular occlusion and a dense neurological deficit may show quite evident improvement over months or years, although the potential for re-learning in nervous circuits may play a part in such prolonged recovery. In more acute circumstances, as for example, the progressive recovery from anaesthesia or the resolution of neurological deficit in recovery from an aneurysm operation by increased blood pressure it is clearly possible for neurones at one moment to be apparently non-functioning and yet under conditions of improved perfusion to return to normal.

The first part of this paper will present some up-to-date view about the recovery of various brain functions after transient ischaemia and in the second, the scientific concepts will be illustrated by a number of clinical examples.

Disturbances of Brain Function During the Development of Brain Ischaemia

Research has been directed towards the analysis of a whole variety of aspects of nervous function during ischaemia. In some ways, however, the most helpful observations in relation to human physiology have arisen from experimental models designed to produce vascular occlusion in defined territories which though severe enough to produce disturbance of function is yet recoverable on reperfusion and bears some resemblance to the characteristics of human cerebral vascular disease. On the other hand, a good deal of research has also employed methods of complete circulatory arrest which have a parallel only in similar circumstances in man and in which the disruption of neural function is more absolute than graded. Nevertheless in certain specific terms, these may be analysed in a helpful fashion.

From the point of view of electrical function, studies have been made of the differential effects of ischaemia on the electrophysiological activity of presynaptic terminals and synaptic transmission[7, 8]. The experimental model of occlusion of the middle cerebral artery has been extensively studied as producing a graded reduction in blood flow in which the relationship between electrical functional failure and recovery on reperfusion may be studied in detail. Most of our personal observations of progression and irreversibility in brain ischaemia have been made in relation to this experimental model. Occlusion of the middle cerebral artery simulates the production of an acute clinical stroke[15, 36] producing an ischaemic lesion whose extent is restricted to known regions of the cerebral hemisphere and which selectively interfers with neural pathways. Such ischaemia is more dense in the centre of the middle cerebral field and progressively diminished towards the collateral bounds of anterior and posterior cerebral arteries[38]. The presence of such a gradient along the sensory motor strip enables assessment of the effects of ischaemia on a peripherally stimulated somatosensory evoked potential whose central transmission is unaffected by middle cerebral occlusion, the thalamic nuclei being outside the area of ischaemia[21, 24]. This technique enables measurement of regional blood flow, as the ischaemia is graded following the occlusion by progressive reduction in blood pressure. Flow is measured by the hydrogen clearance method[27] in the immediate area of the evoked potential electrode, during progressive ischaemia and reperfusion. An apparent threshold of around 16 ml/100 g/min emerged from this study[6].

Using ion sensitive microelectrodes in the extracellular space to measure calcium, potassium and pH[3, 5, 13, 14], it has also been possible to demonstrate similar threshold relationships to flow for each of these variables. While pH clearly changes early in ischaemia in relation to progressive exhaustion of buffering capacity and changes in brain lactate (at or around the level of failure of electrical response), breakdown in ionic homeostasis is associated with very much lower levels of flow. Increase in extracellular potassium concentration and decrease in extracellular calcium concentration did not occur unless local blood flow was reduced to a level of 10 ml/100 g/min. At this point, sudden massive changes in potassium concentration indicated the flux of potassium from the intracellular space, and the rapid disappearance of calcium at only slightly lower levels from the extracellular space has also been shown to be associated with a considerable increase in intracellular calcium, a highly dangerous situation from the point of view of cell metabolism[16].

Biochemical Failure in Ischaemia

The greater part of the information available to us in relation to energy failure has been obtained from experiments involving dense ischaemia produced by total circulatory arrest or bilateral carotid occlusion in small animals such as the rat and gerbil[1, 26, 30]. Under these experimental conditions rapid depletion of high energy organic phosphates with rapid exhaustion of carbohydrates through anaerobic glyocilysis has been demonstrated within seconds of the onset of ischaemia. Brain phosphocreatine and ATP levels have become virtually unmeasurable within a few minutes. The ability to oxidize NADH via the electron transport chain is lost and NADH rapidly accumulates.

Other biochemical disturbances involve amino acid and lipid metabolism. Concentrations of GABA and alanine rise substantially during ischaemia[10]. Possibly more importantly, rapid breakdown occurs in lipid metabolism with release to the extracellular space from membranes, a variety of saturated and unsaturated fatty acids. Most important of these is arachidonic acid leading to the potential degradation cascade with the release of vaso-active components of the cyclooxygenase and lipoxygenase pathways[4, 11]. The disturbance of arachidonic acid metabolism is likely to be of considerable importance for post-ischaemic oedema.

Regrettably, only a few of these phenomena have been studied in detail during re-circulation since the metabolic substrates in particular require sacrifice of the animal for their determination and it is always more difficult to determine the sequences of ischaemic damage and reperfusion under these circumstances.

The Electrical and Biochemical Correlates of Reperfusion: What Will Recover?

The pioneering studies of Dr. Hossmann and his associates some years ago demonstrated clearly and somewhat astonishingly that even after severe brain ischaemia, energy charge potential, high energy organic phosphates and NADH would recover to normal during the recirculation period. Other experiments showed that ADP values virtually unmeasureable during ischaemia would recover to normal by 24 hours even in brain regions which would later develop irreversible neuronal injury. Lactate levels would also recover rapidly but NAD+ levels would generally fail to recover[18, 22].

Immediately after ischaemia there is some evidence of irregular variation in the balance of metabolism and blood flow. Transient periods of increased metabolic rate apparently unassociated with seizures possibly due to a loss of enzymic regulation and uncoupling of oxidative phosphorylation, and certain other areas showing the depressed metabolism in relation to high flow of the state of luxury perfusion.

A number of *in vivo* and *in vitro* experiments have attested to recovery of the cytochrome system during reperfusion and have suggested a return to normal of mitochondrial function. Key experiments by Pulsinelli and Plum[29] have shown animals with excessive ischaemic cerebral acidosis caused either by incomplete brain ischaemia with residual perfusion during the test period, or by high glucose levels before and during ischaemia and have suggested that such tissue acidosis adversely affects mitochondrial function and may irreversibly injure mitochondria. They have given rise to the clear suggestion that the production of high glucose concentration during ischaemia militates against recovery of the tissue.

Key findings again from Hossmann's group relate to the effects of recirculation on protein synthesis. This, virtually abolished during the ischaemic period, remains suppressed for many hours following severe forebrain ischaemia in the rat despite apparently normal RNA synthesis and RNA activity. In the gerbil, five minutes of global ischaemia causes permanent failure of protein synthesis in the hippocampus, suggesting that selective vulnerability may have a link to protein synthesis[25, 29]. Fatty acids return rapidly to normal after resolution of ischaemia, and in the course of this it is possible that reactions could occur leading to the formation not only of vasoactive compounds of the eicosanoid type, but also free radical intermediates.

The question of free radical generation following ischaemia remains complex and uncertain. There is no question that excess free radical concentration in tissue is highly deleterious : the uncontrolled chain reactions to which increased concentration of these compounds could lead normally being quenched by a variety of enzymes, by the vitamins E and C and the tripeptide glutathione.

Regrettably, experimental evidence as to the significance of free radicals remains somewhat in doubt.

The resolution of ionic abnormalities has been studied in a number of experiments. Our own group has shown a return to normal of extracellular potassium activity after one hour of regional ischaemia in the monkey and extracellular pH has been shown to recover to normal within 30 minutes in normoglycaemic rats. Once again, effects of hyperglycaemia have been not only to produce more severe ischaemic and post-ischaemic acidosis but also more prolonged acidosis which may never recover.

Such evidence that we have would suggest that calcium recovers more slowly than potassium but nevertheless will normalize within an hour of recirculation in the primate.

Ischaemic and Post-Ischaemic Oedema

It is of particular interest to study the effects of ischaemia on water movement.

Work from our own laboratory[19, 33] has indicated a definite relationship between the development of cerebral swelling and the intensity of brain ischaemia. In the primate model, for example, assessing brain water content by brain impedance measurements during life together with intracranial pressure changes or by the use of graded density kerosene-bromobenzene columns following the sacrifice of the animal, we found that significant ischaemia was associated at $1^1/2$ hours with an increase in water content in the most densely ischaemic zones but also in the area of the penumbra where potassium movement was absent although evoked responses had been abolished. Movement of water was detected in our preparation when flow fell below $20 \, ml/100 \, g/min$. When ischaemic areas were reperfused after $1^1/2$ hours of the regional ischaemia of the primate model, reperfusion was associated with an increase rather than a decrease in ischaemic oedema. Our observations and those of Hossmann indicate that the degree of post-ischaemic oedema is determined by the initial flow reduction after vascular occlusion in the early phase of brain oedema which we have regarded as cytotoxic rather than vasogenic. Hossmann has also shown that after one hour of complete ischaemia there is a significant increase in tissue osmolality[20] from 308 to 353 mOsm, creating a gradient of about 500 mOsm between brain and blood. We have suggested that this increase in osmolality may be associated with failure of synthesis of large neural transmitter molecules, as a result of which smaller precursors remain free significantly increasing osmolality. The correspondence of the apparent threshold for the initiation of brain oedema and for the failure of the evoked response somewhat supported this hypothesis.

The Relationship of Blood Flow to Tissue Recovery

In almost all preparations in which ischaemia has been induced, reperfusion is accompanied by an initial brief period of hyperaemia followed by a prolonged phase of cerebral hypoperfusion. The complex problem of no-reflow remains to some extent unsettled. Initial experiments of Ames[2] have been regarded as flawed but there seems no doubt that vascular swelling and microaggregation within vessels during the period of ischaemia do result in irregular patchy hyperfusion in certain circumstances. The question of vascular occlusion by physical factors such as endothelial swelling remains controversial, but both increased vascular smooth muscle tone as a result of vasoactive substances released during ischaemia, or permeability changes resulting in pericapillary oedema could produce the increased cerebral vascular resistance demonstrated in certain circumstances in post-ischaemic reperfusion.

Functional Recovery of Tissue Following Ischaemia

In the experimental middle cerebral model, the degree of ischaemia in a given region determined whether or not the evoked response was significantly depressed. When the local blood flow fell below the critical level about 16 ml/100 g/min[6] there was a very severe electrophysiological depression. If this degree of ischaemia was maintained for longer than 15 minutes then it was unlikely that the EP would recover at least over the next one hour period. During this period local blood flow could be observed to be restored but tissue pO_2 remained reduced on average to levels well below control. These experiments suggested distinct post-ischaemic hypometabolism, despite hyperaemic blood flow.

On the other hand if the critical level of ischaemia was not obtained, EP was much less depressed and in a subsequent post clip phase there was a rapid recovery of the EP together with normal flow and a strong suggestion from pO_2 measurements of tissue hyperoxia suggesting the phenomenon of luxury perfusion as described by Lassen[23]. During EP recovery tissue pO_2 tended to be greater than normal probably associated with vasodilatation or local metabolic acidosis[23, 43]. This phenomenon persisted through the post-ischaemic recovery when the tissue flow was not on the average above normal and could perhaps be below it, suggesting that the presence of hyperoxia might well be a necessary condition for altered or complete recovery of function.

Experimental Evidence of Infarct Recovery

Occlusion of the middle cerebral artery in the baboon produces a typical clinical course and a variable hemiparesis. In all instances, however, distinct neurological deficit could be detected and in its most dense portions usually affecting the face, this deficit remained detectable for periods of up to 3 years[36]. However, even

in the most severely affected animals, leg weakness would persist in a severe form for only a day or two after the experimental infarct and would thereafter rapidly recover, and when animals were observed free-walking which was possible in the majority after three or four months, it was almost impossible to detect any weakness of the leg and the animals would leap to a height of 3 or 4 feet without difficulty.

False circling has often been noted in experimental middle cerebral occlusion both in dogs and primates[36, 40]. Later on, however, the circling became a much more casual affair and appeared to rise only from visual inattention. Persistent hemianopia was a characteristic in these animals. The hemiparesis produced by middle cerebral artery occlusion with involvement of the perforating bearing segment was usually fairly dense in the arm. Complete abolition of arm movements even in the first few days after operation was rare and from 10 days onwards most of our animals showed reasonable movement in the proximal muscles and evident recovery in the muscles of the elbow and wrist. The differentiation of stroke density was best performed by considering the time scale of recovery of finger movements and of reaching and placing reactions, but none of the animals ever recovered these movements or a completely normal forelimb. Only one of ten animals recovered to such an extent that it was possible to watch it for a few minutes and be uncertain from its upper limb movements which was the hemiparetic side.

Following such single vessel occlusion in the primate the significant neurological examination did not change form four months after the episode, over an observation period extending to three years.

Human Clinical Correlates

In human terms, however, we are as a rule better able to consider how scientific generalizations fit into the evident pattern of recovery from ischaemia displayed by patients.

Case 1

The first instance cited is recovery from short-term ischaemia which may very well indicate that in man a profound functional loss does indeed follow extremely dense ischaemia after a relatively short time. The case is that of a 36-year-old school teacher who noted over a period of 2 years that his handwriting on the blackboard was becoming smaller and that his right hand was losing to some extent its manipulative dexterity. He was a highly intelligent man, a teacher of mathematics, but put up with the disability for many months before consulting his doctor. His doctor and one of my colleagues, a close friend and an experienced neurologist, thought that the evident mixture of pyramidal and extra-pyramidal signs which he showed, a little stiffness in the arm with positive Hoffman's sign, and slight increase in reflexes in the leg, suggested the early development of idiopathic Parkinson's disease, and indeed he was demonstrated as such to a post-graduate audience to the evident satisfaction of staff and students. CT scan led to a rapid reassessment of the situation. He had a giant middle cerebral aneurysm arising from the trunk of the artery lateral to the deep perforating vessels of the lenticulo-striate group, and proximal to the major trifurcation of the middle cerebral.

Angiography showed that the filling portion of the aneurysm was relatively small in relation to the total mass, the remainder of the sac being occupied by clot. In the course of operation, which was of course conducted with continuous monitoring of electrical function, the middle cerebral artery was occluded both proximal and distal to the aneurysm for a period of 6 minutes and 30 seconds. Within 2 minutes of application of the clips which excluded the lenticulo-striate vessels entirely from the circulation, the evoked response from the middle cerebral cortex disappeared. It remained substantially absent over the next hour although transient middle cerebral occlusion only was employed, but it had to some extent recovered by the return of the patient to the ward. At that time he was aphasic extending the arm and leg and we were deeply concerned about the production of irreversible neurological damage after a period of dense ischaemia in the basal nuclei of no more than 6 minutes and 30 seconds. However, later that day flexion returned to the arm and leg and by that evening he was speaking. Within 48 hours his neurological signs had cleared completely and at the time of his discharge from hospital his IQ was determined as 140, which we believed to be not substantially different from his premorbid IQ although that was not formally assessed. The giant aneurysm was obliterated by the application of a single clip to its neck, and the sac evacuated by the use of the ultrasonic dissection apparatus.

The lesson of this case is that under circumstances of very severe ischaemia, in which one could predict that the residual blood flow would be less than 10 ml/100 g/min in view of the paucity of collateral circulation to the basal nuclei, we have the possibility of apparently complete neurological recovery after a period of 6 minutes and 30 seconds of ischaemia. From our observations in other patients it is possible, that a period of about ten minutes of very dense if not total ischaemia may be sustainable with complete recovery of function as assessed over periods of up to four years, because it is four years since this man had his operation and he remains well. The possible remote effects of such ischaemia must of course be considered and will be dealt with later.

Case 2

The second clinical example concerns the problem of the more gradual recovery from a very dense neurological deficit which one may see in circumstances of brain ischaemia. This woman of 53 had an anterior communicating aneurysm which had bled three times, the last bleed precipitating her into coma. She had no intracerebral haematoma, but she was stable at a level of flexion of both arms with gross weakness of both legs and was operated upon some eight days after haemorrhage by myself, at that time much younger. The aneurysm was successfully occluded but the patient remained comatose with gross leg weakness and poor arm function. The young surgeon discharged her from the hospital to long-term care and took into his consciousness the lesson that grade 4 patients do not do well, a stage one might regard in his acquisition of surgical maturity. CT

scan was not at that time available. Two years later a patient appeared in Outpatients with a folio of cuttings from her local newspaper. She carried a stick because of some residual weakness of one leg; she was voluble but articulate and not dysphasic. The cuttings which she carried documented her miraculous recovery from a period of six months quite unresponsive, in her local hospital. Her final state of capacity showed her to be managing her housework, to be apparently, to her family, slightly more cheerful than before but infinitely more acceptable socially, to be able to remember her shopping lists and to function as a independent housewife without problem.

Once again although it is now nearly 20 years since the operation was performed, and she remains clinically well, we have no absolute assurance of the degree of recovery of all damaged brain, although we have clear evidence that major portions of brain which could not have been functioning at the time of her discharge from the neurosurgical unit, recovered function after a period of six months. It would indeed have been fascinating to have blood flow examinations at that time and although such a striking case has not presented since, a somewhat similar case could be detailed in more recent years.

Case 3

This was also a middle-aged patient, a farmer of 55 years with a severe coma-producing anterior communicating artery aneurysm, operated on in grade 3 with a stable slight weakness of one leg. Serious postoperative vasospasm was associated with fluctuating changes in the electrical conduction in both hemispheres indicating the fluctuation in state of perfusion, and CT scanning showed extensive low density maximal in both frontal lobes, in one instance associated with nonfilling post-operatively of a distal anterior cerebral artery. The aneurysm was successfully occluded from the circulation, and at operation it was not thought that extensive impairment of any peripheral vessel had occurred, but post-operative angiography proved otherwise. His troubled post-operative course went on for several weeks and then, as so many of us have seen, he improved. The CT scan showed undoubted bimedial frontal tissue loss but his hemiparetic signs cleared completely, his intellectual function recovered almost to normal so that he was able to conduct his activities as farm manager without problem and he remains stable some years later. Once again, the clear implication of such a case study is that brain non-functioning immediately after an ischaemic episode, in this instance due to vasospasm compounded by surgical insult, may recover.

Case 4

A fourth case drawn from our records of subarachnoid haemorrhage is a rather younger woman of 35, also with an anterior communicating aneurysm, in this case occluded in grade 2 without neurological deficit. Some days post-operatively, however, hypo-perfusion developed and a fluctuating hemiparesis associated with clear evidence of delay in central conduction indicative of hemispheral ischaemia. Cerebral blood flow showed a reduction in flow in both hemispheres of apparently similar extent to just under 30 ml/100 g/min, but clearly the resolving capacity of the flow determinations were insufficient to pick up the most dense area of ischaemia just below a functional threshold. Again, the induction of hypervolemic hypertension resulted in an improvement in CBF and a resolution of the neurological deficit. Afterwards her blood pressure had to be sustained for nearly two weeks by metaraminol infusion. Whenever the blood pressure was allowed to fall hemiparesis recurred; as the blood pressure was raised, the hemiparesis resolved once more. The capacity, therefore, of brain to retain the ability to recover and yet to be functionally silent for many hours, is adduced by such case.

Case 5

An exactly similar state of affairs applied in a fifth case, a man in his forties who had suffered a severe subarachnoid haemorrhage from a large posterior communicating artery aneurysm round whose neck the anterior choroidal artery was draped. Operated on in grade 2, he developed over the post-operative period a hemiparesis related to spasm of the anterior choroidal artery associated with low density in the deep nuclear distribution of this vessel. Blood flow determined by a 6-channel xenon clearance system showed levels in several counters of below 20 ml/100 g/min, but once again, hypervolemic hypertension using a fluid load with dextran and metaraminol infusion resulted in resolution of the ischaemic deficit.

It is also resulted in the disappearance of the low density area and the patient was discharged without neurological deficit. The metaraminol infusion had to be maintained for five days.

Case 6

A sixth and revealing case is that of a middle-aged woman operated on by a colleague many years before, for an inner third sphenoidal ridge meningioma in the dominant hemisphere. Intense involvement of the middle cerebral artery resulted in a severe post-operative stroke. She was aphasic and hemiplegic. Over the next 3 or 4 years the power in her leg recovered, not unusual one might think in view of the anterior cerebral collateral circulation, but over the next few years, appreciable recovery in her arm also appeared. With the recovery in arm movement she began to recover increasing speech. At first, her speech recovery was in her native language. She was of Central European origin, an immigrant to Britain in the thirties, and it was only many months later that her dysphasia, so evident in English, improved. When I saw her ten years after the operation she was still detectably dysphasic but could manage normal conversation and coped well with her moderate residual hemiparesis. This type of very prolonged recovery, raised speculation about possible retraining mechanisms within the central nervous system but here we are right to doubt such a hypothesis. Consider the evolution of recovery in this particular instance, not an unusual evolution except in the protracted period of time over which it occurred. We have first the recovery of function in areas which could be regarded as closest to increasing collateral blood supply, that is the leg. Then we have recovery of a highly differentiated motor skill, the use of the arm, and finally improvement and function in the most highly developed facet of neurological function which it is possible for us to test, speech. It is of particular interest that the improvement of speech function occurred in that neuronal engram, if we may borrow Penfield's phrase, which was most deeply ingrained, the speech of childhood and adult life. The more lately acquired speech facility, the second language recovered only later. As Gillett (personal communication) points our in a discussion of brain bi-section and personal identity, the loss of extensive portions of brain in hemispherectomy procedures results in severe physical and mental disability. There is no blithe continuation of personal life with largely intact psychological function and no convincing evidence that retraining can occur other than in circumstances of extreme youth. Plasticity of the nervous system, while undoubted in childhood, scarcely extends to the brain of the adult.

Long-Term Problems of Ischaemia

It may seem that in relation to recovery from ischaemia the developments of recent years have indicated an optimistic prospect. A cautionary note, however, must be included in relation to the remote effects of brain damage.

Evidence is emerging from the long-term studies of patients with brain trauma, from head injury or from pugilism, of accelerated dementia in later life. One of the major causes of dementia has been classified as multi-infarct dementia.

The work of Strong and Tomlinson et al.[31] would suggest that while gross neuropathological abnormalities in the area of an ischaemic penumbra are unlikely, some degree of cellular loss appears to be the norm at any rate in the animal species studied by them (the cat). CT scanning of patients with multi-infarct dementia leaves no doubt of cumulative functional loss as a result of repetitive episodes of ischaemia, and mixed pictures of dementia of Alzheimer type and multi-infarct dementia do occur.

It would not be surprising, therefore, if every episode of ischaemia to which the brain was subject resulted in some loss of neuronal reserve and possibly in a contribution to a later dementing process.

The conclusion must be, therefore, that while we have good evidence that transient cerebral ischaemia can apparently recover fully, we must maintain caution in attributing complete safety to techniques which require diminution of the blood supply or the metabolism of the brain. Our task in the coming years must be to analyse more fully the impact on the nervous system with particular reference to its possible cumulative and long-term effects and, of course, to avoid unnecessary manoeuvres as far as possible.

References

1. Abel M, McCandless D (1982) Metabolic profile of hippocampal regions after bilateral ischaemia and recovery. Neurochem Res 7: 789–797
2. Ames A III, Gurian BS (1963) Effects of glucose and oxygen deprivation on function of isolated mammalian retina. J Neurophysiol 26: 617–634
3. Astrup J, Symon L, Branston NM, Lassen NA (1977) Cortical evoked potential and extracellular K^+ and H^+ at critical levels of brain ischaemia. Stroke 8: 51–57
4. Bazan N, Aveldano M, de Caldironi M, Rodriguez de Turco E (1982) Rapid release of free arachidonic acid in the central nervous system due to stimulation or arrests. Lipid Res 20: 523–529
5. Branston NM, Strong AJ, Symon L (1977) Extracellular potassium activity, evoked potential and tissue blood flow, relation-

ship during progressive ischaemia in baboon cerebral cortex. J Neurol Sci 32: 305–321

6. Branston NM, Symon L, Crockard HA, Pásztor E (1974) Relationship between the cortical evoked potential and local cortical blood flow following acute middle cerebral artery occlusion in the baboon. Exp Neurol 45: 195–208

7. Collewijn H, van Harreveld A (1966) Intracellular recording of spinal motoneurones during acute asphyxia. J Physiol 185: 1–14

8. Collewijn H, van Harreveld A (1966) Membrane potential of cerebral cortical cells during spreading depression and asphyxia. Exp Neurol 15: 425–436

9. Dennis C, Kabat H (1939) Behaviour of dogs after complete temporary arrest of the cephalic circulation. Proc Soc Exper Biol Med 40: 559–561

10. Erecinska M, Nelson D, Wilson D, Silver I (1984) Neurotransmitter amino acid levels in the rat brain during ischaemia and reperfusion. Brain Res 304: 9–22

11. Gardiner M, Nilsson B, Rehncrona S, Siesjö B (1981) Free fatty acids in the rat brain in moderate and severe hypoxia. J Neurochem 36: 1500–1505

12. Grenell RG (1946) Central nervous system resistance. I. The effects of temporary arrest of cerebral circulation for a period of two to ten minutes. J Neuropath Exper Neurol 5: 131–154

13. Harris RJ, Symon L, Branston NM, Bayhan M (1981) Changes in extracellular calcium activity in cerebral ischaemia. J Cereb Blood Flow Metabol 1: 203–209

14. Harris RJ, Richards PG, Symon L, Habib A-HA, Rosenstein J (1987) pH, K^+ and pO_2 of the extracellular space during ischaemia of primate cerebral cortex. J Cereb Blood Flow Metabol (submitted)

15. Harvey J, Rasmussen T (1951) Occlusion of the middle cerebral artery. Arch Neurol 66: 20–29

16. Hass WK (1981) Beyond cerebral blood flow, metabolism and ischaemic thresholds: an examination of the role of calcium in the initiation of cerebral infarction. In: Meyer JS et al (eds) Cerebral vascular disease 3. Excerpta Medica, Amsterdam

17. Hossmann KA, Sato K (1971) Effect of ischaemia on the function of the sensorimotor cortex in cat. Electroenceph Clin Neurophysiol 30: 534–545

18. Hossmann KA, Sakaki S, Kimoto K (1977) Cerebral uptake of glucose and oxygen in the cat brain after prolonged ischaemia. Stroke 7: 301–305

19. Hossmann KA, Schuier FJ (1979) The metabolic (cytotoxic) type of brain oedema following middle cerebral artery occlusion in cats. In: Price T, Nelson E (eds) Cerebrovascular diseases. Raven, New York

20. Hossmann KA, Takagi S (1976) Osmolality of brain in cerebral ischaemia. Exp Neurol 51: 124–131

21. Kaplan HA, Ford DH (1966) The brain vascular system. Elsevier, London

22. Kofke W, Nemoto E, Hossmann K-A, Taylor F, Kessler P, Stezoski S (1979) Brain blood flow and metabolism after global ischaemia and post-insult thiopental therapy in monkeys. Stroke 10: 554–559

23. Lassen NA (1966) The luxury-perfusion syndrome and its possible relation to acute metabolic acidosis localized within the brain. Lancet 2: 113–115

24. Lazorthes G, Campan L (1964) La circulation cérébrale. Editions Sandoz, Paris

25. Nowak T, Fried R, Lust D, Passonneau J (1985) Changes in brain energy metabolism and protein synthesis following transient bilateral ischaemia in the gerbil. J Neurochem 44: 487–493

26. Paschen W, Hossmann KA, van den Kerckhoff W (1983) Regional assessment of energy-producing metabolism following prolonged complete ischaemia of cat brain. J Cereb Blood Flow Metab 1: 321–329

27. Pásztor E, Symon L, Dorsch NWC, Branston NM (1973) The hydrogen clearance method in assessment of blood flow in cortex, white matter and deep nuclei of baboons. Stroke 4: 556–567

28. Przybylski A (1971) Activity pattern of visceral cortex neurons during asphyxia. Exp Neurol 32: 12–21

29. Pulsinelli W, Kraig R, Plum F (1985) Hyperglycemia, cerebral acidosis in ischaemic brain damage. In: Plum F, Pulsinelli W (eds) Cerbrovascular diseases, 14th conference. Raven Press, New York, pp 201–206

30. Rehncrona S, Rosen I, Siesjö B (1981) Brain lactic acidosis in ischaemic cell damge: Biochemistry and neurophysiology. J Cereb Blood Flow Metab 1: 297–311

31. Strong A, Tomlinson B, Venables G, Gibson G, Hardy J (1983) The cortical ischaemic penumbra associated with occlusion of the middle cerebral artery in the cat. 2. Studies of histopathology, water content and in vitro neurotransmitter uptake. J Cereb Blood Flow Metabol 3: 97–108

32. Symon L (1961) Studies of leptomeningeal collateral circulation in macacus rhesus. J Physiol 159: 68–86

33. Symon L, Branston NM, Chikovani O (1979) Ischaemic brain oedema following middle cerebral artery occlusion in baboons. Relationship between regional cerebral water content and blood flow at 1–2 hours. Stroke 10: 184–191

34. Symon L, Branston NM, Strong AJ, Hope TD (1977) The concepts of thresholds of ischaemia in relation to brain structure and function. J Clin Pathol 30 [Suppl 11]: 149–154

35. Symon L, Dorsch NWC, Ganz JC (1972) Lactic acid efflux from ischaemic brain. J Neurol Sci 17: 411–418

36. Symon L, Dorsch NWC, Crockard HA (1975) The production and clinical features of a chronic stroke model in experimental primates. Stroke 6: 476–481

37. Symon L, Ganz JC, Dorsch NWC (1972) Experimental studies of hyperaemic phenomena in the cerebral circulation of primates. Brain 95: 265–278

38. Symon L, Pásztor E, Branston NM (1974) The distribution and density of reduced cerebral blood flow following acute middle cerebral artery occlusion: An experimental study by the technique of hydrogen clearance in baboons. Stroke 5: 355–364

39. Takahashi K, Bodsch W, Hossmann KA (1984) Susceptibility of hippocampal protein synthesis to transient forebrain ischaemia of adult and infant gerbil brain. Drugs Dis 1: 72–78

40. Waltz AG (1969) Red venous blood: Occurrence and significance in ischaemic and non-ischaemic cerebral cortex. J Neurosurg 31: 141–148

41. Weinberger LM, Gibbon MH, Gibbon JH Jr (1940) Temporary arrest of the circulation to the central nervous system. 2. Pathologic effects. Arch Neurol Psychiat 43: 961–986

42. Yamaguchi T, Waltz AG, Okazaki H (1971) Hyperaemia and ischaemia in experimental cerebral infarction: correlation of histopathology and regional blood flow. Neurology 21: 565–578

43. Zwetnow NN (1970) Effects of increased cerebrospinal fluid pressure on the blood flow and on the energy metabolism of the brain. Acta Physiol Scand [Suppl] 339: 1–31

Correspondence: Prof. Lindsay Symon, T.D., F.R.C.S., Department of Neurological Surgery, Institute of Neurology, The National Hospital, Queen Square, London WC1N 3BG, U.K.

Acta Neurochirurgica, Suppl. 41, 104–109 (1987)

Vaccinia Growth Factor: Newest Member of the Family of Growth Modulators Which Utilize the Membrane Receptor for EGF

D. R. Twardzik[1], **J. E. Ranchalis**[1], **B. Moss**[2], and **G. J. Todaro**[1]

[1] Oncogen, Seattle, Washington, and [2] Laboratory of Viral Diseases, National Institute of Allergy and Infectious Diseases, NIH, Bethesda, Maryland, U.S.A.

Summary

A computer-aided search for structural homology between epidermal growth factor (EGF), transforming growth factor alpha (TGF-α) and sequences of proteins contained in the Dayhoff data base reveals a statistically significant homology with a peptide predicted to be encoded by an early gene of vaccinia virus (VV), a member of the poxvirus family. Fifteen residues of a 50 amino acid portion of this 140 residue VV polypeptide match residues in TGF-α; after insertion of a single gap, the vaccinia encoded polypeptide shares 19 residues with both EGF and urogastrone. Homologous regions contain six residues that correspond to the six cysteine residues of EGF and TGF-α that form disulphide bond mediated loop structures. A 25,000 M_r (apparent molecular weight) glycosylated polypeptide with the predicted functional activity, competing with EGF for binding to EGF membrane receptors, has been purified to homogeneity from VV infected Cercopithecus monkey kidney cell culture supernatants. This peptide, like both EGF and TGF-α, is a potent mitogen for appropriate target cells. Demonstration of a growth factor encoded by a DNA virus is unprecedented and may expand our understanding of DNA virus-host interactions.

Introduction

We have recently shown that vaccinia virus (VV) infected monkey cells release a growth factor functionally related to epidermal growth factor (EGF) and transforming growth factor alpha (TGF-α)[24]. Both EGF and TGF-α, potent mitogens for a variety of different cells, bind to and phosphorylate in tyrosine residues the 180 kilodalton EGF membrane receptor[3, 18]. Whereas EGF is found primarily in the mouse submaxillary gland[4] and in human urine[11], TGF-α is a product of the transformed phenotype, in particular, retroviral transformed fibroblasts[22–24]. Experiments were initiated following observations from our laboratory[2] and others[1, 17] that EGF and TGF-α (50 and 53 amino acid residues in length respectively) share struc-

tural homology (19 residues in common) with a predicted 19 kilodalton polypeptide of unknown function encoded by an early gene of vaccinia virus. A member of the Poxviridae, vaccinia virus, is a double-stranded DNA virus that replicates within the cytoplasm of infected cells[26, 29]. The 187,000 base-pair genome has a 10,000 base-pair inverted terminal repeat; the gene for the 19 kilodalton polypeptide from the WR strain maps within this inverted terminal repeat at map position 6.54–7.16[28] and is transcribed early in infection[26, 29]. As shown in Fig. 1, this viral encoded 19 kilodalton polypeptide (140 amino acids) is as similar to EGF as is TGF and is referred to in the text as vaccinia growth factor (VGF). After the insertion of a single gap between residues 18 and 19 of rat TGF-α, which correspond to residues 55 and 56 of the VGF sequence, it is possible to align all six cysteines. Conservation is also seen in two tyrosine residues (positions 52 and 76) and several glycines (positions 57 and 75) in the VGF sequence, which is arbitrarily depicted here as starting with residue 38. The first residues in TGF-α (VAL), mEGF (ASN), and hEGF (ASN) correspond to position 1 as was determined by primary amino acid sequence analysis of the native factors. This similar placement of cysteine residues is thought to affect corresponding disulphide bond placement and therefore a similar spatial orientation of the variable intra-loop regions. As a result, a similar secondary structure is proposed. In addition, putative cleavage site and transmembrane sequences suggest other functional similarities between the VV early gene product and the EGF and TGF precursor[8, 10, 13].

Table 1. *Induction of DNA Synthesis by Purified VGF, EGF, and TGF-α*

Growth factor	[^{125}I]IdU incorporated (cpm/plate)	% stimulation
None	876	—
VGF	4,431	406
EGF	5,982	583
TGF-α	4,002	357

VGF was purified as described in text. TGF-α was chemically synthesized utilizing the solid phase system as previously described[21]. EGF was purified from mouse submaxillary gland[5]. Quantitation of EGF equivalents was based on a standard ^{125}[I]-EGF binding competition curve as described. Mitogenesis assays were as described in Materials and Methods. Quiescent cultures of diploid human fibroblasts received 10 ng of purified TGF-α or EGF per ml or the same number of EGF equivalents of Vgf per ml. Values for [^{125}I] iodide-oxyuridine ([^{125}I]IdU) incorporated represent the average of triplicate determinations.

We have previously reported that supernatents of VV infected monkey cells contain an acid and heat stable peptide which completes with EGF for binding to EGF membrane receptors[24]. This 25,000 M_r peptide was partially purified and shown to stimulate DNA synthesis in appropriate serum starved cultures. Release of VGF was proportional to multiplicity of infection and appeared to be the product on an early viral gene (cytosine arabanoside enhanced growth factor production). In another study we also showed that VGF stimulates tyrosine specific phosphorylation of the 180,000 M_r EGF membrane receptor and that antisera prepared against the EGF receptor specifically blocked the observed interaction, *i.e.*, phosphorylation event[12]. At the time we reported these experiments,

similar data was also presented by others[20], including primary sequence data which established VGF as indeed the product of the early VV 19 kilodalton encoded gene. In this report, we describe the purification of VGF to homogeneity and compare the biological activity of the purified molecule to TGF-α and EGF. We also discuss the structural homology between all three members of this family of growth modulating peptides and propose a common basis for the observed equivalent functionality.

Materials and Methods

Cell Culture and Virus

Cercopithecus monkey kidney (BSC-1) cell monolayers were maintained in Eagle's basal medium supplemented with 10% fetal calf serum. VV (strain WR) was grown in HeLa cells and purified by sucrose density gradient sedimentation[16].

BSC-1 cell monolayers were infected with 15 plaque-forming units (pfu) per cell of purified virus, unless otherwise indicated, and incubated at 37 °C with approximately 1 ml of Eagle's basal medium supplemented with 2% fetal calf serum per 2 × 10^6 cells. Mock-infected cells were treated in an identical manner. Cell culture supernatants were clarified by low-speed centrifugation and lyophilized. The residue was then resuspended in 1 M acetic acid and dialyzed extensively against 0.2 M acetic acid. Insoluble material was removed by centrifugation, and the supernatant was lyophilized and resuspended in 1/100th of the original volume of 1 M acetic acid and stored at 4 °C.

Radioreceptor Assay

The binding of ^{125}I-labelled EGF (^{125}I-EGF) to its receptor on monolayers of A 431 cells was modified from that described previously[5]. Cells (1 × 10^3 per well) were fixed on 24-well plates (Linbro, Flow Laboratories) with 10% formalin in phosphate-buffered saline prior to assay. Formalin-fixed cells do not slough off plates as easily as do unfixed cells, and replicate values were thus more consistent.

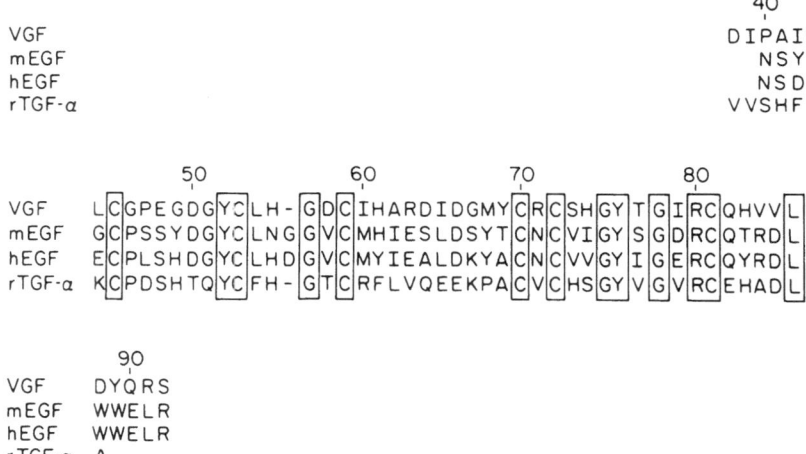

Fig. 1. Comparison of the linear amino acid sequences of VGF with mouse and human EGF and rat TGF-α

Under these assay conditions, ^{125}I-EGF (1×10^{10} cpm/nmol) saturates the binding assay at 3 nM; assays were performed at 10% of the saturation value. TGF and VGF concentrations are expressed as ng equivalents of EGF.

Cellular DNA Synthesis Assay

Diploid human fibroblasts obtained from explants of newborn foreskin were seeded at a density of 3×10^4 cells per well (96 well plates, Nuclon, Roskilde, Denmark) and were grown to confluency in Dulbecco's modified Eagle's medium (GIBCO)/10% newborn calf serum. Cultures were then placed in medium containing 0.2% newborn calf serum, and two days later appropriate growth factors were added. After 8 hours, cultures were labelled with 5-[^{125}I]iodo-2'-deoxyuridine (Amersham, 10 µCi/ml, 5 Ci/mg; 1 Ci = 37 GBq), and the amount of isotope incorporated into trichloroacetate-insoluble material was determined as described[24].

Results

Supernatants were harvested from BSC-1 monolayer cultures 24 hours after infection with the VF strain of vaccinia virus[16]. The media were lyophilized and polypeptides were solubilized with M acetic acid. In a typical experiment, we recovered 14.5 mg of protein containing 551 ng EGF equivalents of VGF (S.A. of 0.038 ng equivalents/µg of protein) from the lyophilizate derived from 1 litre of low serum (0.2%) conditioned media. All VGF and TGF-α activity is expressed as EGF receptor equivalents and was determined using an EGF radioreceptor assay as described in Materials and Methods, using HPLC purified mouse EGF as a reference standard.

As shown in Fig. 2, following chromatography of acid solubilized polypeptides from VV infected cell culture supernatants on Bio-Gel P 100 columns, a major peak of EGF competing activity (fractions 25–35) with an apparent molecular weight of 25,000 is seen. No detectable EGF competing activity is seen on this column in the elution positions of either rat TGF-α or EGF, fractions 55–65 and 75–85 respectively. The activity elutes as a shoulder slightly ahead of the major protein peak. Peak fractions of VGF were pooled (total yield of 0.5 mg of protein containing 1.8 µg EGF equivalents of VGF (S.A. of 3.6 ng EGF equivalents/µg of protein) and chromatographed on a C$_{18}$ µBoundapak column (Fig. 3 A); peptides were eluted with a linear 1 hour gradient of 20 to 60% acetonitile as previously described[8]. A single peak of VGF activity (S.A. 250 ng equivalents/µg for protein) coeluted (30% acetonitrile) with a minor absorbance peak well before the bulk of contaminating serum protein (34–35% acetonitrile). Mouse submaxillary gland EGF elutes from this column at 34% acetonitrile. The peak of EGF-competing activity eluting from the C$_{18}$

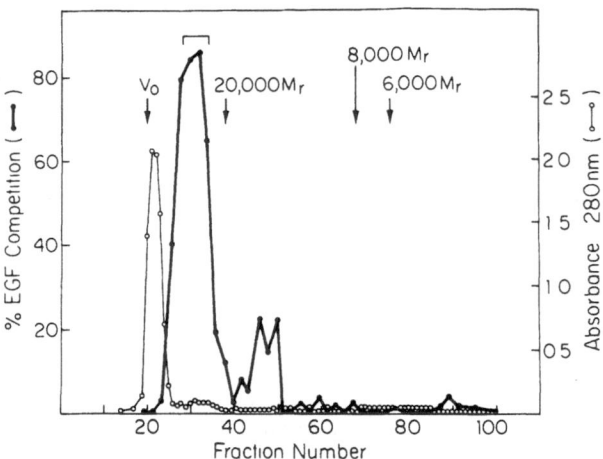

Fig. 2. Gel filtration chromatography of acid solubilized polypeptides derived from VV infected monkey cell (BSC-1) culture supernatents. Lyophilized, acid solubilized, polypeptides from 200 ml of supernatant derived from cells 24 hours after infection with VV (2.9 mg of protein) was dissolved in 7.0 ml of M acetic acid and applied to a Biol-Gel P 100 column equilibrated in 1 M acetic acid. Fractions (3.3 ml) were collected and aliquots of alternate fractions were lyophilized and assayed for EGF competing activity. The column was calibrated with soybean trypsin inhibitor (20 kd); limabean trypsin inhibitor (8 kd); and Insulin (6 kd)

µBondapak column was resolved from contaminating peptides by dilution and rechromatography on the same column utilizing isocratic conditions (Fig. 3 B). All EGF-competing activity coeluted with a major absorbance peak at 214 nM (fraction 14). Purified VGF from this fraction migrated on SDS-polyacrylamide gel electrophoresis as a single, diffusely silver staining band with an apparent molecular weight of 24,000 (J. Cooper, personal communication). The diffuseness of the band after destaining indicated glycosylation. Indeed, experiments were also conducted utilizing ^{3}H-glucosamine labelled monkey cell cultures infected with VV. Utilizing identical purification techniques, the purified, homogeneous VGF band derived from low serum conditioned media in these cultures contained ^{3}H-glucosamine label. The sugar labelled peptide bound to A 431 cells and was completed from the EGF receptor with either cold EGF or VGF. Similar dilution curves suggest identical kinetics.

Discussion

We have recently reported that a polypeptide functionally related to EGF and TGF-α was induced in monkey cells following vaccinia virus infection[24]. Supernatants derived from VV infected monkey cells but not in mock infected cultures contained high levels of an activity

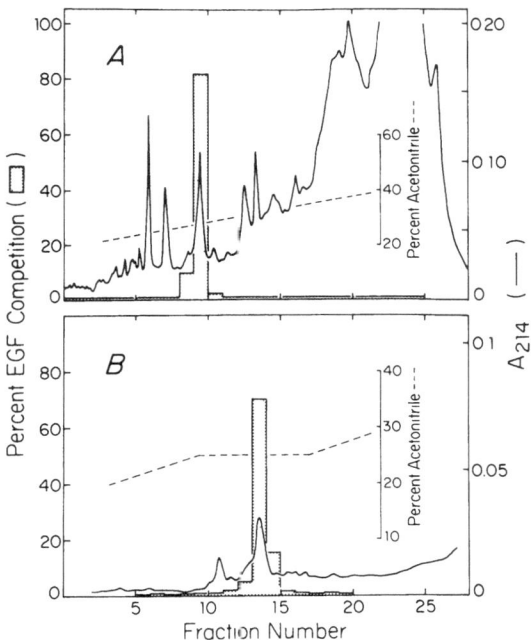

Fig. 3. Purification of VGF by high performance liquid chromatography (HPLC) on μBondapak C_{18} columns. Pooled fractions (25–35) from the gel filtration column were concentrated by vacuum centrifugation, resuspended in 0.05% trifluoroacetic acid (TFA), clarified and injected into a 3.9 mm × 0.5 cm μBondapak C_{18} column (Waters, Milford, MA). Peptides were eluted with a linear 20 to 60% gradient of acetonitrile in 0.05% TFA at a flow rate of 1.0 ml/min at 22 °C (A). Aliquots of each fraction were assayed in a radioreceptor assay for EGF as described[25]. The peptide corresponding to the peak of VGF activity, once located, was selectively collected and diluted with 0.05% TFA and reinjected into a μBondapak column and eluted utilizing isocratic conditions (B). Aliquots of each fraction were assayed and tested for biological activity

that shares biological activity and competes with EGF for binding to EGF membrane receptors. Initial experiments were predicated on the observation by Bloomquist et al.[1], Reisner[17] and our laboratory[2] which described a computer search that revealed a similar pattern of cysteine and glycine residues in EGF, TGF-α, some blood coagulation factors, and the deduced sequence[26] of a 19 kilodalton product of an early gene of unknown function of vaccinia virus (VV). A segment of the putative 140 amino acid viral polypeptide is as similar to EGF as is TGF, with an alignment of cysteine residues that predicts a common secondary structure and thus a presumed similar functionality.

Primary amino acid sequence analysis of VGF[20] shows that the EGF-competing activity released by VV infected monkey cells is indeed the 19 kilodalton product of the early VV gene. The apparent molecular weight of purified VGF is somewhat larger than indicated in previous studies which describe the cell-

free translational product of complimentary RNA to this early VV gene to be 19,000 daltons[26]. The open reading frame of this gene predicts a primary polypeptide of 15,500 M_r, and thus we suspect that post-translational modification such as glyosylation might result in the exaggerated molecular weight. Again, we have shown that ^{3}H-glucosamine labelled VGF is released from VV infected cultures labelled with ^{3}H-glucosamine. In that regard, putative glycosylation sites at residues 34 and 94 are found. A carboxy-terminal analysis will be needed to assess accurately the true size of the processed functional VV 19 kilodalton gene product. Stroobant et al.[20] from amino acid composition and known sequence data calculate a size of 9,084 (77 amino acid residues). However, anomalies in elution properties and exaggerated molecular weights of polypeptides is not uncommon as TGF-α, for example, although eluting from Bio-Gel sizing columns as 10,000 M_r was, upon sequence analysis, shown to be a peptide of only 50 amino acid residues (M_r 7,400)[14, 15].

Brown et al.[2] had previously proposed a model of the VV polypeptide that would be produced by cleavage of the precursor polyprotein at Arg 43 and Arg 90 thus generating the release of a soluble polypeptide 47 residues in length similar in size to EGF and the small form of TGF-α. Amino-terminal analysis of VGF, however, indicates cleavage between residues 19 and 20, alanine and aspartic acid, thus suggesting a unique cleavage site and corresponding peptidase with such specificty. In this regard, the amino-terminal processing of the TGF-a precursor is also unique as the native 50 amino acid peptide is preceded and followed by Ala Ala Val Val and Val Val Ala Ala sequences, respectively[8, 13].

The remarkable similarity in secondary structure of all three polypeptide members of this family of growth factors is visualized in Fig. 4. Using the proposed structure of EGF as the prototype[19], both TGF-α and VGF easily adapt to the EGF conformation. Three major loops or domains are formed as dictated by the three disulphide bridges. Of interest, is the observation that structural homology in most of the intraloop regions is poor. Interpretations of some studies utilizing synthethic peptides corresponding to various loop regions of EGF suggest a particular region may be involved in membrane receptor interactions. This does not, however, seem to be the case for TGF-α as we have tried unsuccessfully to demonstrate a synthethic peptide with the aforementioned activity (DRT). Most likely, we suggest that the receptor recognition site will be the result of the cumulative interaction of the entire

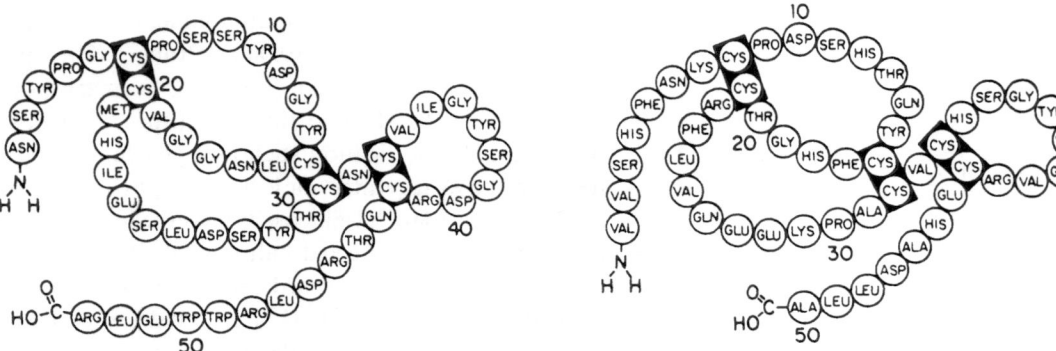

a. Epidermal Growth Factor   b. Transforming Growth Factor - alpha

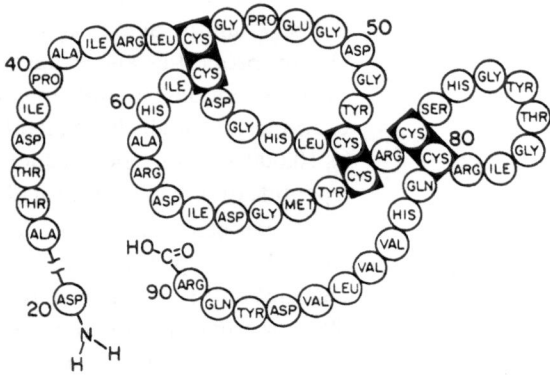

c. Vaccinia Growth Factor

Fig. 4. Models proposing secondary structure of (*a*) EGF (Savage et al., 1972); (*b*) TGF-α; and (*c*) VGF

molecule (in toto) as manifested in the formation of three-dimensional "domains".

The encoding of a growth factor by VV or any other DNA virus is unprecedented. A member of the RNA tumour virus family of viruses, Simian Sarcoma Virus (SSV), however appears to have acquired a gene with homology to a chain of platelet derived growth factor (PDGF)[27]. In addition, it has also now been found that some RNA viruses also contain genes encoding not only growth factors but also fragments of their receptors. No doubt the placement of cysteine residues in VGF has been conserved by VV throughout many years of evolution and is not a chance event. Experiments to probe the function of VGF in the life cycle of this large member of the pox family are in progress. The release of high levels of a potent mitogen immediately after viral infection may be responsible for the hyperplastic response commonly associated with some pox virus infections, and thus may indicate that the association of growth factors with some DNA viruses may represent a more generalized phenomenon than previously thought.

Acknowledgements

We wish to thank Dr. Anthony Purchio for constant encouragement, and Nancy Olfs for her assistance in the preparation of this manuscript.

References

1. Blomquist MC, Hunt LJ, Barker WC (1984) Vaccinia virus 19-kilodalton protein: relationship to several mammalian proteins, including two growth factors. Proc Natl Acad Sci USA 81: 7363–7367
2. Brown JP, Twardzik Dr, Marquardt H et al (1985) Vaccinia virus encodes a polypeptide homologous to epidermal growth factor and transforming growth factor. Nature 313: 491–492
3. Carpenter G, Cohen S (1979) Epidermal growth factor. Annu Rev Biochem 48: 193–216
4. Cohen S (1962) Isolation of a mouse submaxillary gland protein accelerating incisor eruption and eyelid opening in the newborn animal. J Biol Chem 237: 1555–1561
5. Cohen S, Carpenter G (1975) Human epidermal growth factor: isolation and chemical and biological properties. Proc Natl Acad Sci USA 72: 1317–1321
6. Cohen S, Carpenter G, King L Jr (1980) Epidermal growth factor-receptor-protein kinase interactions. Co-purification of

receptor and epidermal growth factor-enhanced phosphorylation activity. J Biol Chem 255: 4834–4842

7. DeLarco JE, Todaro GJ (1978) Growth factors from murine sarcoma virustransformed cells. Proc Natl Acad Sci USA 75: 4001–4005

8. Derynck R, Roberts AB, Winkler ME et al (1984) Human transforming growth factor-alpha: precursor structure and expression in E. coli. Cell 38: 287–297

9. Doolittle RF, Hunkapiller MW, Hood LE et al (1984) Simian sarcoma virus oncogene v-sis, is derived from the gene (or genes) encoding a platelet-derived growth factor. Science 221: 275–277

10. Gray A, Dull TJ, Ullrich A (1983) Nucleotide sequence of epidermal growth factor cDNA predicts a 128,000-molecular weight protein precursor. Nature 303: 722–725

11. Gregory H (1975) Isolation and structure of urogastrone and its relationship to epidermal growth factor. Nature 257: 325–327

12. King CS, Cooper JA, Moss B, Twardzik DR (1986) Vaccinia virus growth factor stimulates tyrosine protein kinase activity of A 431 cell epidermal growth factor receptors. Mol Cell Biol 6: 332–335

13. Lee DC, Rose TM, Webb NR et al (1985) Cloning and sequence analysis of a cDNA for rat transforming growth factor-alpha. Nature 313: 489–491

14. Marquardt H, Hunkapiller MW, Hood LE et al (1983) Transforming growth factors produced by retrovirus-transformed rodent fibroblasts and human melanoma cells: amino acid sequence homology with epidermal growth factor. Proc Natl Acad Sci USA 80: 4684–4688

15. Marquardt H, Hunkapiller MW, Hood LE et al (1984) Rat transforming growth factor type 1: structure and relation to epidermal growth factor. Science 223: 1079–1082

16. Moss B (1985) Replication of poxvirus. In: Fields BN, Chanock RM, Roizman B (eds) Human viral disease. Raven Press, New York, pp 685–703

17. Reisner AH (1985) Similarity between the vaccinia virus 19 K early protein and epidermal growth factor. Nature 313: 801–803

18. Reynolds FH Jr, Todaro GJ, Fryling C et al (1981) Human transforming growth factors induce tyrosin phosphorylation of EGF receptors. Nature 292: 259–262

19. Savage CR Jr, Inagami R, Cohen S (1972) The primary structure of epidermal growth factor. J Biol Chem 247: 7612–7621

20. Stroobant P, Rice AP, Gullick WJ et al (1985) Purification and characterization of vaccinia virus growth factor. Cell 42: 383–393

21. Tam JP, Marquardt H, Rosberger DF, Wang TW, Todaro GJ (1984) Synthesis of biologicall active rat transforming growth factor I. Nature 309: 376–378

22. Todaro GJ, Fryling CM, DeLarco JE (1980) Transformation induced by Abelson murine leukemia virus involves production of a polypeptide growth factor. Proc Natl Acad Sci USA 77: 5258–5262

23. Todaro GJ, Marquardt H, Twardzik DR et al (1982) Transforming growth factors produced by viral-transformed and human tumour cells. In: Weinstein IB, Vogel HJ (eds) Genes and proteins in oncogenesis. Academic Press, New York, pp 165–182

24. Twardzik DR, Brown JP, Ranchalis et al (1985) Vaccinia virus-infected cells release a novel polypeptide functionally related to transforming and epidermal growth factors. Proc Natl Acad Sci USA 82: 5300–5304

25. Twardzik DR, Todaro GJ, Marquardt H et al (1982) Transformation induced by Abelson murine leukemia virus involves production of a polypeptide growth factor. Science 216: 894–897

26. Venkatesan S, Gershowi TZ, Moss B (1982) Complete nucleotide sequences of two adjacent early vaccinia virus genes located within the inverted terminal repetition. J Virol 44: 637–646

27. Waterfield MD, Scrace GT, Whittle N et al (1983) Platelet-derived growth factor is structurally related to the putative transforming protein p28sis of simian sarcoma virus. Nature 304: 35–39

28. Wittek R, Cooper JA, Banbosa E, Moss B (1980) Expression of the vaccinia virus genome: analysis and mapping of mRNAs encoded within the inverted terminal repetition. Cell 21: 487–493

29. Weir JP, Bajszar G, Moss B (1982) Mapping of the vaccinia virus thymidine kinase gene by marker rescue and by cell-free translation of selected mRNA. Proc Natl Acad Sci USA 79: 1210–1214

Correspondence: D. R. Twardzik, M.D., Oncogen, 3005 First Avenue, Seattle, WA 98121, U.S.A.

Acta Neurochirurgica, Suppl. 41, 110–117 (1987)

Oncogenes Related to Growth Factor Receptors

K. Toyoshima

Institute of Medical Science, University of Tokyo, Tokyo, Japan

Summary

Recent progress on analysis of functions of proto-oncogenes was discussed with the protein-tyrosine kinase group as an example. Many proto-oncogenes encode proteins related to regulation of cell growth and differentiation. Among those, the largest group is the protein kinase super family which includes growth factor receptors, non-receptor type protein-tyrosine kinases and serine-threonine kinases. They can be divided into subgroups according to their polypeptide and genomic structures. Members of each subgroup appeared to be originated from each single ancestral gene during evolutionary process even within the protein kinase super family.

Keywords: Oncogene; growth factor receptor; non-receptor type protein-tyrosine kinase; *erbB/src*.

Initial Studies of Oncogenes

The oncogene study was initiated by analysis of the retroviral gene responsible for controlling cellular transformed phenotypes using Rous sarcoma virus. The first two isolates of ts-mutants were temperature-sensitive not only for initiation of transformation, but also for maintenance of the transformed state. The foci produced by these mutants at the non-restrictive temperature disappeared when the cultures were shifted up to 40.5 °C[26]. The following year, Martin (1970) showed that the phenotype of cells transformed by his ts-mutants could be regulated reversibly between the transformed and normal states by shift of the culture-temperature between the permissive (36 °C) and the non-permissive (41 °C) temperatures[15].

On the other hand, isolation of non-conditional transformation defective mutants[26, 28] led to the molecular biological analysis of the transforming gene of Rous sarcoma virus[8]. Steheling et al.[25] prepared complementary DNA specific to the transforming gene of Rous sarcoma Virus (cDNA^{src}) with the aid of reverse transcriptase in virions. All strains of Rous sarcoma virus contained a sequence hybridizable to cDNA^{src} now called the viral *src* gene (v-*src*), but there were other acutely oncogenic retroviruses whose genomes did not hybridize with cDNA^{src}. More importantly they found that DNA hybridizable to cDNA^{src} was present in normal avian cellular DNAs and now it is called the proto-*src* gene (c-*src*). This finding was further extended to the discoveries of a similar sequence in various vertebrates[24].

These findings facilitated search for oncogenes, eventually leading to the discovery of many viral and cellular oncogenes related with acutely oncogenic retroviruses. These discoveries unified the concepts of the mechanisms of chemical and viral oncogenesis as the activation of cellular oncogenes.

Relation Between Oncogene Products and Growth Promoting Elements

Molecular studies on retroviral oncogenes revealed that they can be divided into at least 4 families (Tables 1 and 2): (i) The kinase superfamily: Nearly half of the well-characterized oncogenes belong to this superfamily. Their products have domains with significant amino acid sequence homology similar to that of pp 60^{src}, an oncogene product of Rous sarcoma virus. Such a homologous structure is responsible for protein-tyrosine kinase activity or serine-threonine kinase activity. (ii) *sis* representing the growth factor related genes: Oncogenes belong to this family are still few. (iii) The *ras* family: Their gene products have highly related sequences and show guanine nucleotide binding activity. In addition, the products of this family are suggested to be related functionally with G-proteins. (iv)

Table 1. *Proto-Oncogenes of the Kinase Super-Family*

Non-receptor type	src family:	src, yes, fgr, syn, lyn, lck
	fps/fes	
	abl	
Receptor type	erbB family:	erbB/EGF-R, erbB-2/neu
	ros family:	ros-1, ros-2, IGF-R?
	fms family:	fms/CSF-1, PDGF-R
	unclassified:	met/mil, sea
Serine-threonine kinase		raf, A-raf, mos

Table 2. *Proto-Oncogenes Other Than the Kinase Super-Family*

Growth factors	sis, Blym/transferrin?, TGF
GTP-binding proteins	ras family: H-ras, K-ras, N-ras
Nuclear proteins	myc family: C-myc, N-myc, R-myc
	unclassified: myb, fos, ets, p53, erbA
Unclassified	hst, lca (transfection)
	bcl-1, bcl-2, tcl-1, tcl-2 (translocation)
	rel, jun (retroviral transduction)
	int-1, int-2 (insertional activation)

( ) Method used for detection of oncogenes.

The *myc* and other genes encoding nuclear proteins: The gene products have DNA binding activity and are expected to regulate gene expression. All of these oncogenes related somehow to the elements for control of cell growth.

As mentioned above, the largest group is the kinase superfamily. Of particular interest is the fact that protein-tyrosine kinase activity is also associated with several receptors for polypeptide growth factors, such as epidermal growth factor (EGF)[4], platelet-derived growth factor (PDGF)[16], insulin[12] and insulin-like growth factor I[19].

In fact, recent analysis of the v-*erbB* gene, and the EGF receptor gene indicated that the v-*erbB* gene is a truncated version of the EGF receptor gene[7, 23, 35]. More recently, the cellular homologue of the v-*fms* gene, an oncogene of the McDonough strain of feline sarcome virus, was found to encode a receptor for the mononuclear phagocyte growth factor, CSF-1[23]. These findings, together with the extensive identity of the amino acid sequence of the v-*sis* gene product and PDGF[6, 30], suggest an attractive hypothesis that deregulated expression of the growth factor receptor is important in the neoplastic process.

v-*erbB*, the Oncogene of Avian Erythroblastosis

The R strain of avian erythroblastosis virus (AEV), was originally isolated by Engelbreth-Holm (1931) and a single virus could cause both erythroblastosis and sarcoma *in vivo*, and could transform erythroid cells as well as fibroblastic cells *in vitro*.

Two independently isolated strains of AEV, the R (or ES 4) strain and H strain, have been extensively characterized both virologically and molecularly. The genome of AEV-R carries two cell-derived sequences, termed v-*erbA* and v-*erbB*. In contrast, the genome of AEV-H consists of the v-*erbB* gene and the viral *gag* and *pol* genes, but not the *erbA* gene[32]. Since the H strain is capable of inducing both erythroleukemia and fibrosarcomas *in vivo*, the v-*erbB* gene is evidently responsible for the induction of these tumours. The same conclusion was deduced by analysing the transforming capacity of deletion mutants of the *erbA* gene and *erbB* gene in AEV-ES 4[9].

However, the localization of the v-*erbB* gene product was clarified only recently. In transformed cells this second protein was found as a membrane-associated glycoprotein[11]. Erythroblastosis caused by *erbB* alone had a slightly longer latent period and progress of the disease appeared to be slightly milder than that caused by AEV-R or ES 4. The function of *erbA* is still not fully understood, but is supposed to strengthen the function of *erbB* in some way and to block the differentiation properties of cells transformed by *erbB* alone[9].

In the case of AEV ES 4, the *erbB* product, p 62.5, was glycosylated to gp 68 and further to gp 74. The former is located in the Golgi area and the latter at the cell membrane. In chicken cells infected by a ts mutant for transformation, glycosylation of the *erbB* product from gp 68 to gp 72 was inhibited concomitantly with its translocation from the Golgi to the cell membrane at the non-permissive temperature, suggesting the importance of the localization of this glycoprotein at the cell membrane[3].

Structures of the *erbB* Gene and Its Product

The nucleotide sequence of the *erbB* gene of AEV-H was determined. The total open reading frame is 1812 nucleotides long and codes for a polypeptide with a calculated molecular weight of 67,638, which is consistent with the molecular weight estimated by gel electrophoresis (67,000 for non-glycosylated and 72,000 for glycosylated protein) (Fig. 1).

Unexpectedly, the middle portion of the deduced amino acid sequence shows significant homology

Similarities between p67^{erbB} and p60^{src}

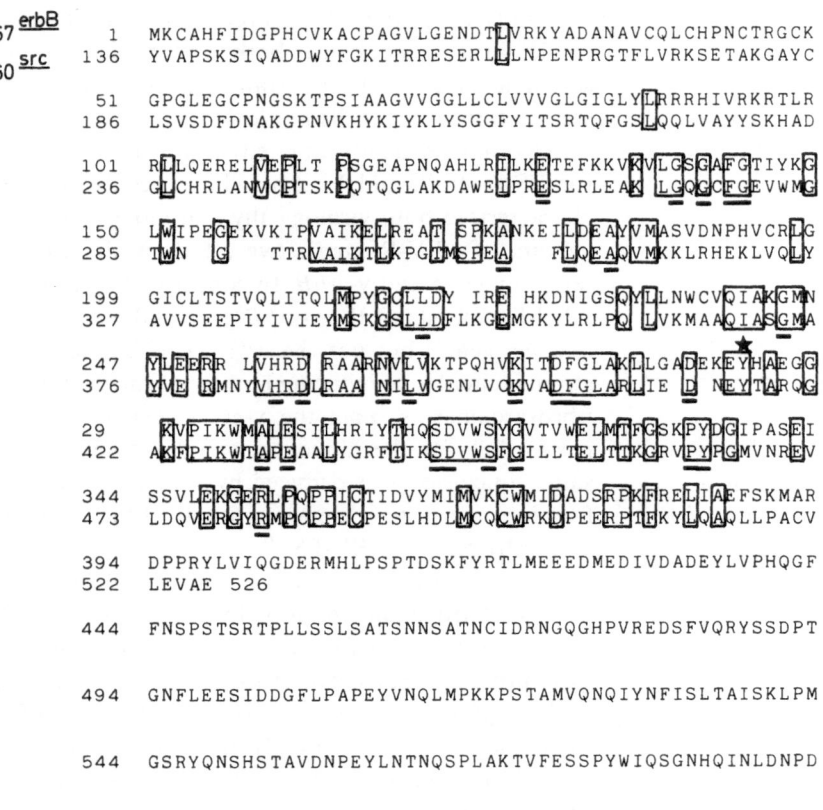

Fig. 1. Simliarities between p67^{erbB} and p60^{src}. Matched amino acids are boxed. Numbers indicate positions of amino acid in each product

(38%) with that of the carboxy half of p60^{src}. All the amino acids conserved in common in the various products of the *src* family were also conserved in the *erbB* product[35]. The conserved portion in the *src* family was named the S domain by our group[35], but later the same part was called the kinase domain[18] and we now feel that the latter is the better name.

td-130, which has lost the capacity to transform erythroblasts but still retains the capacity to transform fibroblasts, has a deletion of 169 bp immediately after the region homologous to the kinase domain and shows frame shift after the deletion. There is a termination codon soon after the frame shift and td-130 has a truncated product of about 47,000 daltons, indicating a decrease in molecular weight of about 20,000[35]. From these data together with the fact that the kinase domain is highly conserved in the *src* family, the kinease domain appears to be essential for the transforming capacity of the gene product of this family.

The amino terminal portion of the *erbB* protein contains a 21 residue hydrophobic amino acid stretch, which could be a transmembrane portion, and three possible N-glycosylation sites near the amino terminus. The product is actually attached to the membrane and glycosylated.

At the beginning of 1984, Downward et al. found that 6 of 14 polypeptides generated by tryptic digestion of the human epidermal growth factor receptor (EGFR) showed striking similarity in its amino acid sequences to portions of the *erbB* protein deduced from its nucleotide sequence. Soon after this finding, complementary DNA to mRNA of EGFR was cloned and sequenced[28]. The amino acid sequence deduced from their nucleotide sequence data indicated that the *erbB* protein corresponds roughly to the carboxyl half of EGFR. The former does not have a receptor domain, but has some gylcosylation sites, a transmembrane stretch, a kinase domain and an E domain corresponding to the carboxyl terminus of EGFR. However, a short stretch of the amino acid sequence at the extreme carboxy end of EGFR is not present in the *erbB* product. These results strongly suggest that the normal counterpart of the *erbB* protein is EGFR, the small difference in amino acid sequences may be explained by the difference in the species, chicken (*erbB*) and human (EGFR), from which these genes were obtained.

Amplification of the EGF Receptor Gene in Human Cancer

High similarity between EGF receptor and the *erbB* product led to analysis of DNA from a human epidermal cancer cell line, A 431, which was known to express EGF receptor at high level. By using complementary DNA (cDNA) to v-*erbB*, 50- to 60-fold amplification of the gene was observed. In addition, amplification of EGF receptor was also observed in 10 of 12 squamous cell carcinoma cell lines[34]. Synthesis of EGF receptor is also higher in the cells with amplified EGF receptor gene. In contrast, incidence of amplification of EGF receptor gene in primary tumours is not very high, suggesting that epidermal carcinomas with amplified EGF receptor gene might readily adapt to culture conditions.

Recent findings revealed that amplification and partial deletion of the EGF receptor gene was also present in some human glioblastoma lines maintained in nude mice (Shibuya, personal communication).

Discovery of the c-*erbB*-2 Gene

As described in the previous chapter, A 431 cells express a high level of EGF receptor as a consequence of amplification of the EGF receptor gene. Southern blotting analysis with v-*erbB* probe under relaxed conditions detected a few bands, which were not amplified in A 431 cells, besides amplified bands derived from the EGF receptor gene (Fig. 2). A genomic clone containing a part of this novel v-*erbB* related sequence was isolated from a placenta gene library. The nucleotide sequence of a part of this clone revealed that this gene may encode a protein-tyrosine kinase whose amino acid sequence is highly homologous to the kinase domain of *erbB* product. This gene was named c-*erbB*-2 and was shown to be conserved among various vertebrates like other proto-oncogenes[21].

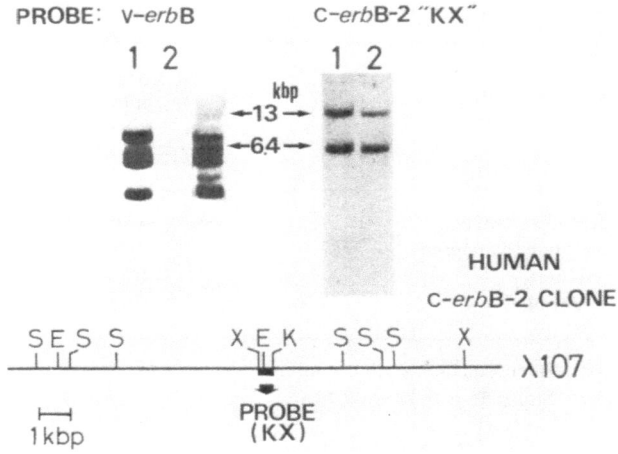

Fig. 2. Southern blotting of *erbB*/EGF receptor gene and c-*erbB*-2. *1* DNA from A 431, human vulva carcinoma cell line. *2* DNA from human normal placenta and long exposure of lane 2 in the left panel. v-*erbB* cDNA to v-*erbB*, c-*erbB*-2 "KX" was obtained from λ 107 as shown in lower panel

Amplification of c-*erbB*-2 in Human Cancers

Amplification of c-*erbB*-2 was detected with specific probe, in a human adenocarcinoma of the salivary gland, but not in a human squamous cell carcinoma[21]. Later, the DNAs from 101 fresh human malignant tumours were tested with the 3.0 kbp HindIII-KpnI fragment of cDNA to human c-*erbB*-2 mRNA. Amplification of the c-*erbB*-2 gene was observed in two of 9 stomach adenocarcinomas, one of 43-kidney adenocarcinomas and two of 10 breast adenocarcinomas, but not in any other tumours, including 8 squamous cell carcinomas, in one of which amplification of the EGF-R gene was observed[37]. These results are in clear contrast to results on the c-*erbB*-1/EGF-R gene, which is amplified in some squamous cell carcinomas, but rarely in adenocarcinomas[34] (Table 3).

The amplification of a v-*erbB* related gene in a human mammary carcinoma reported by King et al.[13] may also be indicate the involvement of c-*erbB*-2 with human cancer.

The c-*erbB*-2 locus was mapped on human chromosome 17 at q21 by using two independent methods, hybridization of sorted chromosomes with K-X probe and *in situ* hybridization experiments on metaphase-chromosome spreads with pCER 217 plasmid DNA[10].

Structure of the c-*erbB*-2 Gene Product

For examination of the function of the c-*erbB*-2 product, the nucleotide sequence of the total coding

Table 3. *Amplification and Over-Expression of* erbB *and* erbB-2

	Gene amplification		Immuno-fluorescent staining with anti-	
	erbB	*erbB-2*	*erbB*	*erbB-2*
Adeno carcinoma	1[a]/66	6/66	1/32	6[b]/32
Squamous carcinoma	2/16	0/16	1/1	0/1
Sarcoma	0/23	0/23	0/4	0/4
Leukemia	0/10	0/10	—	—

[a] Adenosquamous cell carcinoma.
[b] Stomach 4/10, breast 2/16.

region of c-*erbB*-2 was determined. A cDNA library was constructed with poly(A) RNA prepared from MKN-7, a human stomach cancer cell line, which is known to express c-*erbB*-2 mRNA at high level and then representative clones were compared by their restriction maps and partial sequencing.

A nucleotide sequence of 4.480 bp was obtained from representative clones from the cDNA library. The longest open reading frame in this sequence is 3,765 bp, encoding a protein with 1,255 amino acid residues and a calculated molecular weight of 137,895. The total amino acid sequence deduced from the nucleotide sequence is compared with that of EGF-R in Fig. 3[33].

Approximately in the middle of this sequence, there is a stretch of 22 hydrophobic amino acid residues corresponding to the transmembrane domain of EGF-

R. Toward the amino terminus from the hydrophobic stretch is a 653 amino acid stretch including putative signal peptides at the extreme amino terminus. This stretch includes 2 cysteine clusters, correspond to those of EGF-R, which are considered to be very important for maintaining the structure of the receptor. Altogether, this stretch shows 44% homology in amino acid sequence with that of EGF-R and is thought to be the extracellular domain. Toward the carboxyl terminus from the transmembrane domain, there is a stretch of 580 amino acids. It includes the kinase domain and the carboxyl-terminus domain which show 82% and 32% homology in amino acid sequence, respectively, with the corresponding parts of EGF-R. The total amino acid sequence of c-*erbB*-2 also shows 88% homology with that of the *neu* gene[2], which was isolated by the transfection method from a rat neuro/glioblastoma[20], suggesting that these two genes are the human and rat counterparts of the same gene.

The c-*erbB*-2 Product and Its Function

Antibodies were raised against synthetic peptides corresponding to the 14 carboxyl terminal amino acids deduced from the c-*erbB*-2 nucleotide sequence. The lysate of MKN-7 cells labelled with ^{35}S-methionine was mixed with the antibody and the resultant immunoprecipitate was analysed by gel electrophoresis. A 185 kD protein was immunoprecipitated by the antibody, but not by antibody preincubated with the peptide used for immunization. This protein was not detectable in a lysate of HeLa cells in which c-*erbB*-2 mRNA is expressed at only low level. On the other hand, the EGF-R molecule of 170 kD was immunoprecipitated with anti EGF-R antibody from both MKN-7 and HeLa cell lysates. These results suggest that the 185 kD protein is the product of the c-*erbB*-2 gene[1].

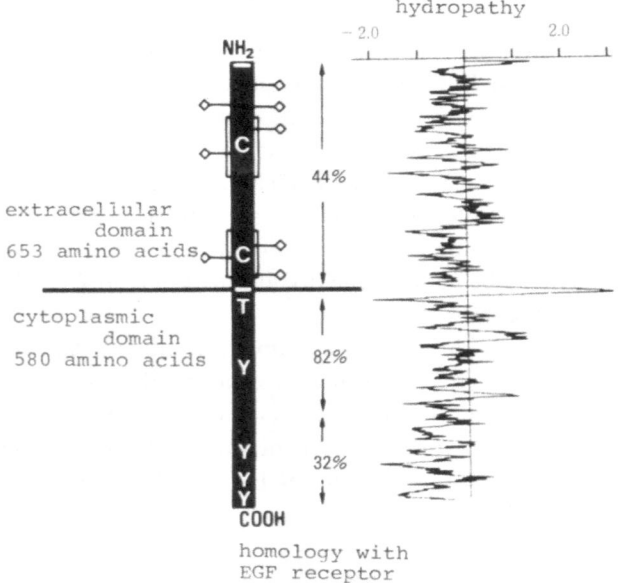

Fig. 3. Model of c-*erbB*-2 product. [C] Cysteine cluster; *T* threonine; Y tyrosine; ⟡—possible glyosylation sites

The c-erbB-2 product has the kinase domain and 4 tyrosine residues that are common to various oncogene products of the tyrosine kinase group and are frequently phosphorylated when the immunoprecipitate is incubated with ^{32}P-ATP. As expected, the c-erbB-2 product in the immunoprecipitate was phosphorylated on its incubation with ATP. EGF did not show appreciable binding to this product. In spite of extensive studies, the ligand for this receptor has not yet been found, but the observation of gene-amplification and higher expression of c-erbB-2 in adenocarcinomas suggests that this receptor acts principally in glandular cells.

Structural Resemblance Among 6 Non-Receptor Type Protein Kinases

The non-receptor type protein kinase closely related to src and yes have molecular weight of about 60,000. Although amino terminal portion of the c-fgr product is still not known, it is also expected to have similar molecular configuration (Fig. 4). Including fgr, there are six genes coding for related kinases. Similarity of these protein in the kinase domain is around 80% and is much higher than the similarity to other protein tyrosine kinases ranging between 40 to 60%. For N-terminus, glysine at the second is conserved. This glysine in the src product is known to be myristylated after removal of methionine and contributes for trapping the protein to the inside of cell membrane[5]. Thus, all of these protein kinases are assumed to localize inside of the cell membrane and may contribute for signal transmission. Approximately 80 amino acids after glysine at the 2nd position show the lowest homology of about 20 to 40% among these five kinases. Since other portions including C-terminus retain high

homology, this unique portion may possibly assign specificity of each enzyme.

The levels of transcription of these genes are different in different tissues and organs. For example, syn expression in relatively high in embryonal brain and in placenta[22] whereas lyn expression is higher in embryonal liver and in placenta[36] and fgr expresses at appreciable level only in B cell line transformed by EB virus (Nishizawa et al., 1986).

Different expression of these genes in different tissues and unique structures toward the amino terminus of each gene product suggest that each gene may play a unique role in each of these specified tissues.

Comparison of Genomic Structures of Protein-Tyrosine Kinases

The erbB and c-erbB-2 genes code for receptor type proteins which show homology of amino acid sequence along entire products. In addition, these two genes have identical splicing junctions at all sites so far sequenced[21] (Haley et al., personal communication). Similarly, 6 non-receptor type protein tyrosine kinases described in the previous chapter show homology of amino acid sequence along entire products. The genes encoding these proteins also have structural similarity and the splicing junctions so far sequenced are completely identical. However, the splicing junctions of c-erbB and src, the representative of receptor type and non-receptor type tyrosine kinasae genes, are completely different. No site is identical even within the kinase domain in which amino acid sequences show appreciable homology up to 40%.

These results suggest that c-erbB and c-erbB-2 were originated by duplication of a single ancestor gene and

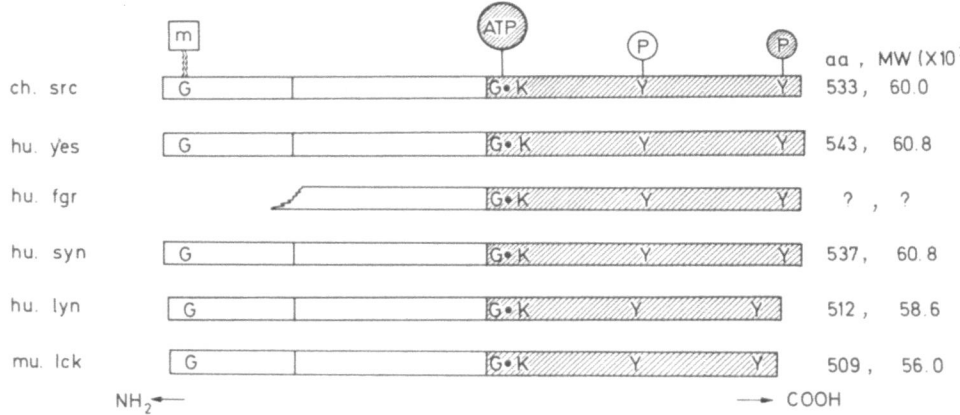

Fig. 4. Comparison of 6 non-receptor type proteins. m Meristilation site; ATP binding site; P phosphorylation site; aa number of amino acids

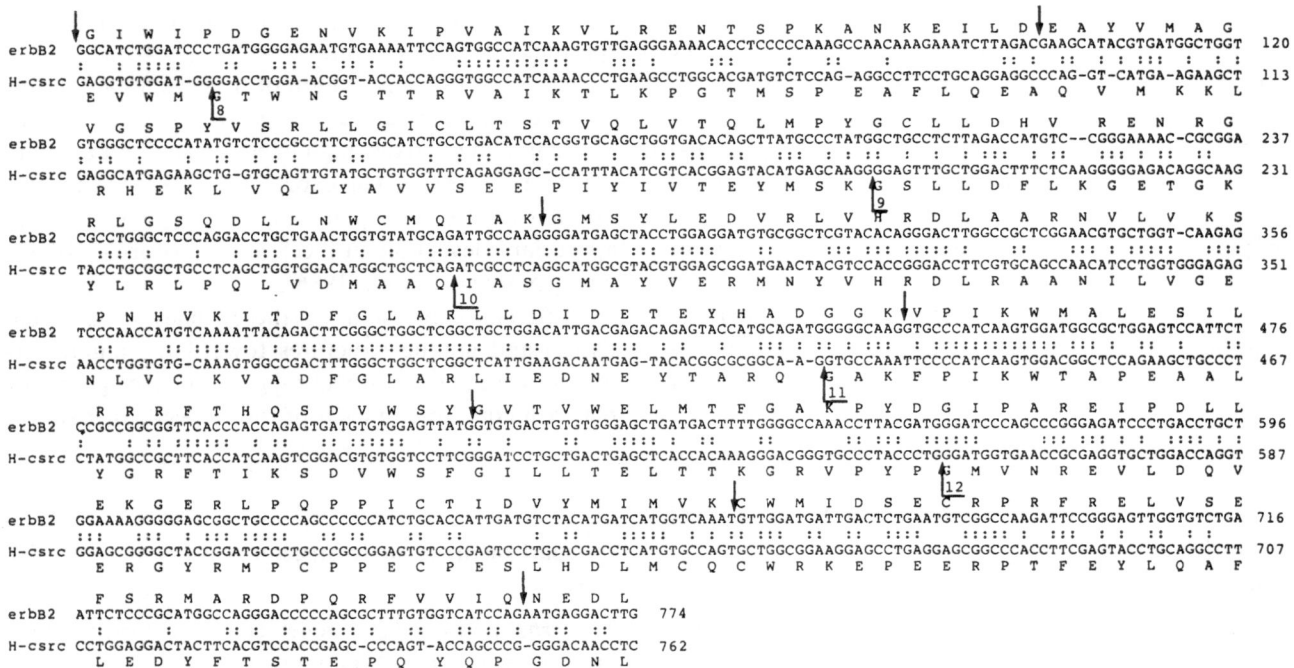

Fig. 5. Comparison of splicing sites in the kinase domain between human c-*erbB*-2 and human c-*src*. Arrows: splicing sites. Numbers with arrows: exon number of c-*src*. Numbers on right end: nucleotide in the kinase domain

that the 6 genes coding for non-receptor type tyrosine kinases were also descendants of another ancestor gene. Likewise, various tyrosine kinases with different structures appeared to have different splicing sites probably because the originated from different ancestor genes. These differences may be related to their specific functions and to their specific target cells for oncogenic transformation.

References

1. Akiyama T, Sudo C, Ogawara H et al (1986) The product of the human c-*erbB*-2 gene: a 185-kilodalton glycoprotein with tyrosine kinase activity. Science 232: 1644–1646

2. Bargmann CI, Hung M-C, Weinberg RA (1986) The neu oncogene encodes an epidermal growth factor receptor-related protein. Nature 319: 226–230

3. Beug A, Hayman MJ (1984) Temperature-sensitive mutants of avian erythroblastosis virus: Surface expression of the *erbB* product correlates with transformation. Cell 36: 963–972

4. Cohen S, Carpenter G, King L Jr (1980) Epidermal growth factor receptor protein kinase interactions. Co-purification of receptor and epidermal growth factor enhanced phosphorylation activity. J Biol Chem 225: 4834–4842

5. Cross FR, Garba EA, Pellman D et al (1984) A short sequence in the p60*src* N terminus is required for p60*src* myristylation and membrane association and for cell transformation. Mol Cell Biol 4: 1834–1842

6. Doolittle RF, Hunkapiller MW, Hood LE et al (1983) Simian sarcoma virus oncogene, v-*sis*, is derived from the gene (or genes) encoding a platelet derived growth factor. Science 221: 275–277

7. Downward J, Yarden Y, Mayed E et al (1984) Close similarity of epidermal growth factor receptor and v-*erbB* oncogene protein sequence. Nature 307: 521–527

8. Duesberg PH, Vogt PK (1973) RNA species obtained from clonal lines of avian sarcoma and avian leukosis virus. Virology 54: 207–219

9. Frykberg L, Palmieri S, Beug H et al (1983) Transforming capacities of avian erythroblastosis virus mutants deleted in the *erbA* and *erbB* oncogenes. Cell 32: 227–238

10. Fukushige S, Matsubara K, Yoshida MC et al (1986) Localization of a novel v-*erb B* related gene, c-*erb B*-2, on human chromosome 17 and its amplification in a gastric cancer cell line. Mol Cell Biol 6: 955–958

11. Hayman M, Ramsay G, Savin K et al (1983) Identification and characterization of the avian erythroblastosis virus *erbB* gene product as a membrane glycoprotein. Cell 32: 579–588

12. Kasuga M, Zick Y, Blithe DL et al (1982) Insulin stimulated tyrosine phosphorylation of the insulin receptor in a cell-free system. Nature 298: 667–669

13. King CR, Kraus MH, Aaronson SA (1985) Amplification of a novel v-*erbB*-related gene in a human mammary carcinoma. Science 229: 974–976

14. Marth JD, Peet R, Krebs EG et al (1985) A lymphocyte-specific protein-tyrosine kinase gene is rearranged and overexpressed in the murine T cell lymphoma LSTRA. Cell 43: 393–404

15. Martin GS (1970) Rous sarcoma virus: a function required for the maintenance of the transformed state. Nature 227: 1021–1023

16. Nishimura J, Huang JS, Devel TF (1982) Platelet derived growth factor stimulated tyrosine specific protein kinase activity in Swiss mouse 3T3 cell membranes. Proc Natl Acad Sci USA 79: 4303–4307

17. Nishizawa M, Semba K, Yoshida MC et al (1986) Structure,

expression and chromosomal location of the human c-*fgr* gene. Mol Cell Biol 6: 511–517

18. Privalsky ML, Ralston R, Bishop JM (1984) The membrane glycoprotein encoded by the retroviral oncogene v-*erbB* is structurally related to tyrosine-specific protein kinases. Proc Natl Acad Sci USA 81: 704–709

19. Rubin JB, Shia MA Pilch PF (1983) Stimulation of tyrosine-specific phosphorylation *in vitro* of insulin-like growth factor-I. Nature 325: 438–440

20. Schechter AL, Stern DF, Vaidyanathan L et al (1984) The neu oncogene: An *erbB*-related gene encoding a 185,000-Mr tumor antigen. Nature 312: 513–516

21. Semba K, Kamata N. Kawano H et al (1985) A new *erbB* related proto-oncogene, c-*erbB*-2, is distinct from the c-*erbB*-1/EGF receptor gene and is amplified in a human salivary adenocarcinoma. Proc Natl Acad Sci USA 82: 6497–6501

22. Semba K, Nishizawa M, Miyajima N et al (1986) *yes*-related proto-oncogene, *syn*, belongs to the protein-tyrosine kinase family. Proc Natl Acad Sci USA 83: 5459–5463

23. Sherr CJ, Rettenmeir CW, Sacca R et al (1985) The c-*fms* proto-oncogene product is related to the receptor for the mononuclear phagocyte factor, CSF-1. Cell 41: 665–676

24. Spector D, Varmus HE, Bishop JM (1978) Nucleotide sequences related to the transforming gene of avian sarcoma virus are present in the DNA of uninfected vertebrates. Proc Natl Acad Sci USA 75: 4102–4106

25. Stehelin D, Varmus HE, Bishop MJ et al (1976) DNA related to the transforming genes of avian sarcoma virus is present in normal avian DNA. Nature 260: 170–173

26. Toyoshima K, Friis FR, Vogt PK (1970) The reproductive and cell-transforming capacities of avian sarcoma virus B 77: Inactivation with UV light. Virology 42: 163–170

27. Toyoshima K, Vogt PK (1969) Temperature sensitive mutants of an avian sarcoma virus. Virology 39: 930–931

28. Ullich A, Coussens L, Hayflick JS et al (1984) Human epidermal growth factor receptor cDNA sequence and aberrant expression of the amplified gene in A 431 epidermal carcinoma cells. Nature 309: 418–425

29. Vogt PK (1971) Spontaneous segregation of nontransforming viruses from cloned sarcoma viruses. Virology 46: 939–946

30. Voronova AF, Sefton BM (1986) Expression of a new tyrosine protein kinase is stimulated by retrovirus promoter insertion. Nature 319: 682–685

31. Waterfield MD, Scrace T, Whittle N et al (1983) Platelet-derived growth factor is structurally related to the putative transforming proteins p28^{sis} of simian sarcoma virus. Nature 304: 35–39

32. Yamamoto T, Hihara H, Nishida T et al (1983) A new avian erythroblastosis virus, AEV-H, carried *erbB* gene responsible for the induction of both erythroblastosis and sarcoma. Cell 34: 225–232

33. Yamamoto T, Ikawa S, Akiyama K et al (1986) Similarity of protein encoded by the human c-*erbB*-2 gene to epidermal growth factor receptor. Nature 319: 230–234

34. Yamamoto T, Kamata N, Kawano H et al (1986) High incidence of amplification of the EGF receptor gene in human squamous carcinoma cell line. Cancer Res 46: 414–416

35. Yamamoto T, Nishida T, Miyajima N et al (1983) The *erbB* gene of avian erythroblastosis virus is a member of *src* gene family. Cell 35: 71–78

36. Yamanashi Y, Fukushige S, Semba K et al (1987) The *yes*-related cellular gene, *lyn*, encodes a possible tyrosine kinase similar to p 56^{lck}. Mol Cell Biol 237–243

37. Yokota J, Yamamoto T, Toyoshima K et al (1986) Frequent amplification of the c-*erbB*-2 oncogene in human adenocarcinomas. Lancet i: 765–766

Correspondence: K. Toyoshima, M.D., Institute of Medical Sciences, University of Tokyo, 4-6-1 Shiroganedai Minato-ku, Tokyo 108, Japan.

Acta Neurochirurgica, Suppl. 41, 118–125 (1987)

Growth Factors and Receptors of Lymphocytes

T. Kinashi, Y. Noma, Ch. Azuma, Sh. Kondo, T. Tanabe, M. Konishi, M. Kinoshita, M. Takahashi, Y. Saito, T. Ogura, H. Sabe, A. Shimizu, Y. Yaoita, and **T. Honjo**

Department of Medical Chemistry, Kyoto University Faculty of Medicine, Kyoto, Japan

Summary

To understand molecular mechanisms of clonal expansion of lymphocytes we have isolated cDNA clones for two lymphokines, interleukins (IL) 4 and 5 that induce proliferation and maturation of B-lymphocytes. Structures of IL-4 and IL-5 revealed a remote homology with other lymphokines such as IL-3 and γ-interferon. IL-4 and IL-5 were shown to affect not only B-lymphocates but also T-lymphocytes and several other cells derived from bone marrow stem cells. We have also studied the structure and function of the IL-2 receptor: our focus was the molecular basis for the high and low affinity states of the receptor encoded by the identical cDNA. We propose the affinity conversion model that the high-affinity state of the IL-2 receptor is a ternary complex of IL-2, the IL-2 receptor, and a postulated "converter protein", which is fewer in number than the receptors.

Keywords: Interleukin-4; interleukin-5; nucleotide sequence; interleukin-2 receptor; converter protein; affinity conversion model.

Introduction

The central nervous system is characteristic of enormous diversity and the complex network. Generation of the complex neuron network is one of the most fascinating problems in modern biology. The degree of diversity and complexity of the immune system is as high as that of the central nervous system. The recent advances of molecular genetics have greatly increased our knowledge of molecular mechanisms to create immune diversity[1, 2]. In essence the generation of immune diversity employs the Darwinian principle; generation of genetic variation and selection of lymphocyte clones. It is now clear that somatic recombination and mutation of the immunoglobulin genes contribute to the amplification of germ line immune diversity. Since such genetic events produce random variants, it is essential to select clones which are useful to the defense system of the living organism.

Lymphocytes stimulated by antigens proliferate specifically because only antigen-stimulated lymphocytes secrete growth factors (lymphokines) and express their receptors on their surface.

The general picture of creation of immune diversity may have important implications for understanding the mechanism of generation of nervous diversity. Selective survival of neurons was suggested to take place during embryogenesis of neuron junctions (reviewed in 3). Recent studies have shown that the immune system has a direct influence and relationship with the nervous system: (a) astrocytes proliferate in the presence of a lymphokine, interleukin (IL)-1[4], and (b) proliferation and maturation of oligodendrological cells is stimulated by IL-2[5].

We will review our recent studies on lymphokines and their receptors to explain how clonal selection is accomplished in the immune system. First, we will describe our results on cloning two major lymphokines which were originally thought to be involved solely in B lymphocyte maturation but actually involved in maturation and proliferation of many other cell lineages derived from the bone marrow stem cells[6, 7]. Secondly, we will briefly summarize our recent studies on the IL-2 receptor, specifically asking a question what is the molecular basis for the high and low affinity species of this receptor[8].

Molecular Cloning of cDNAs Encoding Interleukins 4 and 5

Proliferation and maturation of antigen-stimulated B cells are regulated by several soluble factors derived from macrophages and T cells[9, 10]. These soluble factors were functionally divided into two groups; one called B-cell growth factor (BCGF) which was thought to be involved in B-cell proliferation and the other called B-cell differentiation factor (BCDF) which would be responsible for maturation of activated B cells into immunoglobulin-secreting cells[11-14]. However, the distinct classification and identification of the B-cell factors should be re-examined after the recent cloning of cDNA IgG_1 induction factor (IL-4) from 2.19 T-cell line[6]. This factor induces not only an elevated IgG_1 response in B cells activated by lipopolysaccharides but also hyper I a expression in B cells. IL-4 is identical to B-cell stimulating factor-l (BSF-l) which induces DNA synthesis given together with anti-IgM antibodies. Furthermore, this lymphokine has growth factor activities for both T and mast cells[6, 15]. Another well-characterized B-cell factor is T-cell replacing factor (TRF)[16-18]. TRF secreted by a murine T-cell hybridoma B151K12 was defined by two activities: (a) induction of IgM secretion by BCL_1 leukaemic B-cell line and (b) induction of secondary antidinitrophenol (DNP) IgG synthesis *in vitro* by DNP-primed B cells. Although TRF of B151K12 was previously classified into BCDF, a partially purified preparation of TRF was suggested to have the BCGFII activity. The identity of TRF with BCGFII was proved by its cDNA cloning and the name IL-5 was proposed for this lymphokine[7]. We will review our studies on the structure and multiple activities of these lymphokines, IL-4 and IL-5.

Construction of pSP6K Vector Libraries

Since these factors were scarce and antibodies were not available, we needed a good expression vector to screen the cDNA library for these factors. Some of us developed a new expression vector containing the SP6 promoter which allowed synthesis of a few micrograms of RNA by *in vitro* trancription with a specific RNA polymerase[19, 20]. A basic strategy of cDNA library construction is similar to that described by Okayama and Berg[21] except that the *Hind*III linker segment containing the SV40 promoter sequence is replaced by another *Hind*III linker containing the SP6 promoter sequence. To obtain the SP6 promoter *Hind*III linker, the plasmid pSP62-PL was modified to pSP62-K2. The preparation of the linker and construction of the cDNA library are shown schematically in Fig. 1. We refer to this vector as the pSP6K system.

The pSP6K vector has several advantages over the pCDVl vector. (a) Only a small amount (1–3 µg) of pSP6K DNA is required to be transcribed into mRNA, which is directly injected into *Xenopus* oocyte. (b) Usually the amount of mRNA obtained by *in vitro* transcription with SP6 RNA polymerase is equal to that of template DNA. This procedure avoids all the problems associated with DNA transfection to COS cells. (c) The pSP6K system is able to produce more concentrated supernatants than the pCDVl system. Using cloned human IL-2 cDNA inserted into the pSP6K vector, we were able to obtain $1.6 \times 10^5 \, U \, ml^{-1}$ of the IL-2 activity in oocyte culture media. On the other hand, murine IL-2 cDNA in the pCDVl vector produced $6.9 \times 10^3 \, U \, ml^{-1}$ of the IL-2 activity in culture supernatants of COS cells transfected with this clone[22]. Supernatants of *Xenopus* oocyte culture are free from mycoplasma contamination and contain few other proteins. We could concentrate oocyte supernatants a further 10-fold. In fact, previous cloning with the pCDVl vector was carried out by pooling 40–50 clones, but we could easily detect IL-4 and IL-5 activities by pooling the total library (45,000 clones) for poly(A)$^+$ mRNA.

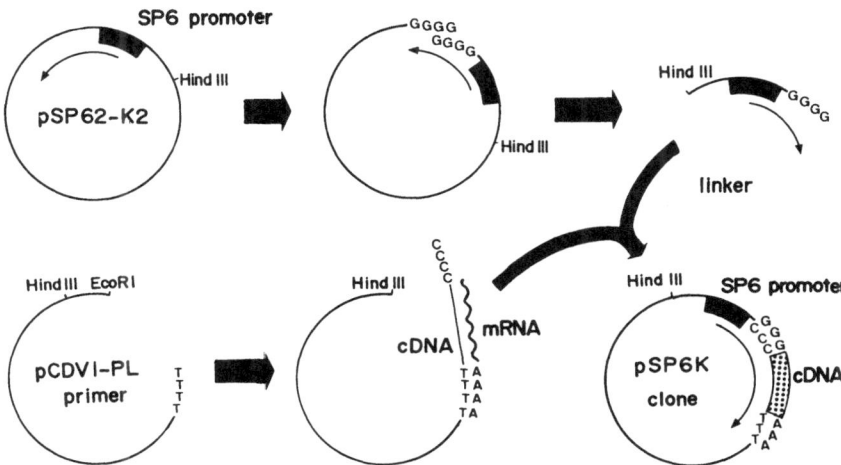

Fig. 1. Strategy for construction of pSP6K library. Closed and dotted columns show SP6 promoter and double-stranded cDNA, respectively. Arrows indicate the direction of transcription by SP6 polymerase

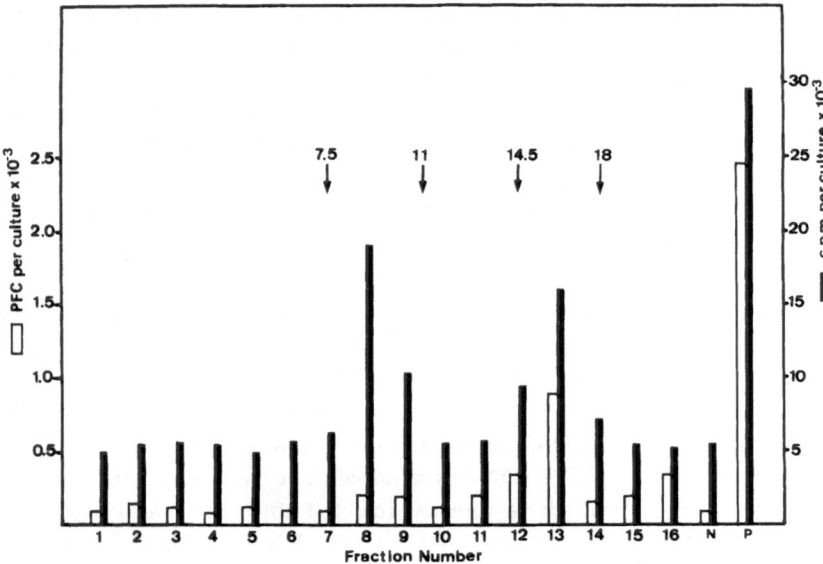

Fig. 2. The effects on BCL₁ cells of translational products directed by size-fractionated mRNA of 2.19 T cell line. Poly(A)⁺ RNA was obtained from concanavalin A-activated 2.19 T cell line, and fractionated into 16 fractions by centrifugation in a 5–22% sucrose gradient, at 36,000 r.p.m. for 15 h. An aliquot of each fraction was injected into *Xenopus* oocytes. Incubation media were collected after incubation for 36 h at 20 °C and added to the assay medium at the concentration of 1%. ³H-thymidine incorporation (closed bars) and anti-IgM plaque-formation response (open bars) of BCL₁ cells were measured as described in Table 2. P, positive control where culture supernatants of B 151 hybridoma were added to the medium at the concentration of 50%. N, negative control where phosphate buffer was injected into *Xenopous* oocytes instead of mRNA. Arrows indicate sizes of mRNA in Svedberg units

Identification of mRNA for IL-4 and IL-5 in 2.19 T Cell RNA

2.19 T cell line was established as a source of IL-4. Poly(A)⁺ RNA of 2.19 T cells was fractionated by sucrose density gradient centrifugation and an aliquot of each fraction was injected into *Xenopus* oocytes to test the presence of mRNA encoding IL-4 and IL-5 by measuring proliferation stimulation and IgM plaque-formation of BCL₁ cells. The BCL₁ proliferation activity was found in translation products directed by 9 S and 16 S mRNAs as shown in Fig. 2. In contrast, IgM plaque-forming activity was found only in products of 16 S mRNA. 9 S mRNA corresponds to the size of mRNA for IL-4[6]. Translation products directed by 15–18 S mRNA of B151K12 cells were shown to exert the TRF activity[7]. We therefore concluded that 16 S mRNA of 2.19 T cells encoded another factor IL-5 which had the TRF activity as well as BCGFII defined by the growth stimulation of BCL₁.

Construction of SP6K Library of cDNA of 2.19 Cell mRNA

We have constructed pSP6K libraries of cDNA for total poly(A)⁺ RNA from the 2.19 T-cell line. This library contained 4.5×10^4 independent cDNA clones. DNA of the library was cleaved with several restriction enzymes to linearize plasmid DNA, and capped RNA was synthesized using SP6 RNA polymerase. The synthesized RNA was injected into *Xenopus* oocytes and translation products secreted into the incubation

media were assayed for IL-4 and IL-5. As shown in Table 1, the library for total poly(A)⁺ RNA of 2.19 T cells significantly augmented the IgG₁ response when template DNA was cleaved with *Sal*I. Digestion of library DNA with *Pst*I or *Sac*I completely abolished the activity, indicating that cDNA encoding the IL-4 contains both *Pst*I and *Sac*I sites in the structural sequence.

Table 1. *IgG1-Inducing Activities Directed by pSP6K cDNA Clones*

Clones used as template	IgG1 plaque-forming cells per culture
ᵃ *Sal*I-digested library	558
*Sac*I-digested library	12
*Pst*I-digested library	18
Phosphate buffer	138
Positive control	858
ᵇ pSP6K-280	140
pSP6KmIL4-374	1,680
Phosphate buffer	84
Positive control	2,520

In experiments *a* and *b*, oocyte culture media were added to concentrations of 20 and 2%, respectively, to the B-cell cultures with LPS. Positive control is the 2.19 T-cell culture supernatant, which was added to a concentration of 5%. The transcripts (10–20 ng per 70 nl) were injected into one *Xenopus* oocyte. Ten oocytes for each sample of RNA were incubated in 100 µl Barth's medium at 20 °C for 36 h. Incubation medium was gently collected and aliquots were added to LPS-containing cultures of mouse spleen cells (200 µl per well). The numbers of cells secreting immunoglobulin of different isotypes were estimated in the protein A plaque assay, using subclass-specific developing antisera. The responses shown are from days 5 and 6.

Table 2. *TRF and BCGFII Activities on BCL₁ of Supernatants of Cell Culture and mRNA-Injected Oocytes*

Cells of supernatants, or templates of mRNA injected into oocytes	IgM plaque-forming cells/culture	Thymidine incorporation (cpm)/culture
[a] *Sac*I-digested library	500	8,396
*Sal*I-digested library	799	14,342
Phosphate buffer	563	6,732
[b] pSP6K 16	199	1,728
pSP6K 18	233	1,969
pSP6K-mTRF 23	3,132	17,970

Supernatants of oocytes injected with mRNA were added to the culture medium of BCL₁ cells to 5%. After two days culture, the protein A plaque assay was used to determine the numbers of IgM secreting cells. BCL₁ proliferation was determined by ^{3}H-thymidine uptake for the last six hours of a 2-day culture.

Translation products of *Sal*I-cleaved template were active for the proliferation and IgM plaque-forming cells of BCL₁ whereas those from *Sac*I-digested templates did not have any of them, indicating that cDNAs encoding BCGFII and TRF contain *Sac*I sites in the structural sequences (Table 2).

Isolation and Structure of IL-4 and IL-5 cDNA Clones

The pSP6K cDNA libarary of 2.19 T cell mRNA was divided into pools, each considering of 3000 clones. We assayed each pool for IL-4 and IL-5 production. After repeated subpooling steps of positive pools, we identified positive clones for IL-4 or IL-5. IL-4 and IL-5 cDNA clones were designated as pSP6KmIL4-374 and pSP6K-mTRF23, respectively. The nucleotide sequences of the inserts of pSP6KmIL4-374 and pSK6K-mTRF23 were determined[6, 7]. The longest open reading frame of IL-4 deduced polypeptide of Mr 15,836, which contained three N-glycosylation sites and a 15-residue hydrphobic region at the N terminus. The hydrophobic signal sequences could be extended to residue 24 although it was interrupted by two hydrophilic residues (Glu 16 and Arg 19). The Mr of the secreted core polypeptide would be 14,137–13,012, which is reasonable for a glycosylated protein of Mr 20,000. The longest open reading frame contains *Sac*I and *Pst*I sites in the coding region

The reading frame of IL-5 encodes a polypeptide of 133 amino acids with three possible N-glycosylation sites. Twenty-one hydrophobic amino acids are located at the NH₂-terminus. The secreted core polypeptide has a Mr of 12,300, which is reasonable to behave as a TRF molecule of Mr 18,000 after glycosylation.

Biological Activities

The cDNA for the IgGl induction factor was the first B-cell factor to be cloned. We have further tested this lymphokine for several other activities that are thought to be associated with IgGl induction factor. First, we asked whether the product of a single gene can induce IgGl production and reduce IgG3 and IgG2b secretion. Supernatants of oocytes injected with pSP6KmIl4-374mRNA or from the 2.19 T-cell line were added to LPS cultures and the IgG subclass responses were assessed. In the absence of added supernatants, LPS stimulated high IgM, IgG2b and IgG3 responses, but a low IgGl response[23]. Addition of either of the supernatants resulted in a dose-dependent increase in IgGl secretion and a concomitant decrease in the level of IgG2b and IgG3 isotypes. No significant effect by the supernatants on the IgM response was observed. Supernatants derived from oocytes injected with mRNA from a negative clone (pSP6K280) had no effect on any isotype. These data demonstrate that the changes in the IgGl, IgG2b and IgG3 responses are caused by the products of a single gene, as suggested previously[23].

Based on the biochemical similarities between the IgGl induction factor and the B-cell stimulating factor 1 (BSF-1)[24], some of us have previously suggested that the two factors are identical[23]. The latter factor is defined by its capacity to act as a co-stimulator of DNA synthesis together with anti-immunoglobulin. Supernatants from the 2.19 T-cell line or translation products of the cDNA clone pSP6KmIL4-374 were able to synergize with anti-immunoglobulin in inducing DNA synthesis in B lymphocytes. Supernatants from oocytes injected with phosphate buffer were inactive in this assay.

Another function that has been assigned to BSF-l is its capacity to induce I a expression in resting B lymphocytes[25]. We show here that the product of the cloned gene as well as supernatants from the 2.19 T-cell line induced an almost threefold increase in the amount of I a expressed on B lymphocytes without any significant increase in cell size.

We have tested various biological activites of recombinant TRF (rTRF), *i.e.*, the oocyte translation product of pSP6K-mTRF23. First, we tested rTRF for the other activity of TRF, namely, stimulation of the secondary anti-DNP IgG response. Mice were immunized with DNP-keyhole limpet haemocyanin (KLH), and splenic B cells were purified 6–8 weeks later for the assay of the anti-DNP IgG plaque-forming cells. The assay revealed that rTRF stimulated DNP-primed B cells in the presence of antigen (DNP) and augmented anti-DNP IgG response. rTRF also induced IgM synthesis in *in vivo* activated B cell blasts (data not shown). These properties of rTRF completely agreed with the criteria of TRF dervied from B 151[18].

In the course of the screening, we found that pools which showed the stimulating activity of IgM plaque-forming cells of BCL$_1$ cells, had always the stimulation activity of the BCL$_1$ proliferation. The final clone mTRF23 also showed these two activities. These results unequivocally demonstrate that the TRF and BCGFII activities are catalyzed by a single molecule. rTRF had no activity of IL-1, IL-2, IL-3 or BSF-I. rTRF induced augmentation of ^{3}H-thymidine uptake in thymocytes in conjunction with suboptimal doses of ConA and IL-2. Sanderson *et al.* have recently suggested that BCGFII is identical to eosinophil differentiation factor.

Because of the complexity of the process of B-cell activation, the different steps involved have previously been considered to be independently controlled events. Several experimental systems meant to identify and to characterize the factors regulating each of these events were constructed. As a consequence many B-cell-specific factors have been reported and classified as growth, differentiation or maturation factors. The fact that a single molecular species (IL-4 or IL-5) can participate in several activation processes has implications for our view of B-cell activation. It may be an oversimplified view of this dynamic process to consider the B-cell response as an ordered series of independent events, each of them regulated independently by the action of different controlling elements.

Molecular Basis for Two Different Affinity States of the Interleukin 2 Receptor: Affinity Conversion Model

Antigen-specific clonal proliferation of lymphocytes is an important selection mechanism of lymphocytes

bearing specific antigen receptors. The physiological proliferation of T cells requires interaction between the humoral growth factor, IL-2 and its cell-surface receptor[26, 27]. Normal resting T cells, which do not express the IL-2 receptor, fail to receive growth signals even in the presence of IL-2. Antigenic stimulation of T cells induces transient expression of the IL-2 receptor, indicating that regulated expression of the IL-2 receptor is the molecular basis for positive selection of antigen-specific T-cell clones[28, 29]. IL-2 receptors are of two distinct populations with high and low affinities[30]. Physiological signals by IL-2 seem to be mediated through binding of the growth factor to the high-affinity receptor[31].

Studies of human and murine IL-2 receptors started with cloning of their cDNA[31–36]. The IL-2 receptors expressed on non-lymphoid cells by cDNA transfection all have low affinity and are inactive in signal transmission[37, 38]. The IL-2 receptor proteins sythesized in non-lymphoid cells when analyzed by immunoprecipitation are indistinguishable from those in T cells[37]. We have established, using cDNA transfection, an IL-2 dependent mouse T-cell line, CT/hR-l, that expresses the human IL-2 receptor as well as the murine receptor[39]. Human IL-2 functions with both human and murine receptors, so the activity of human IL-2 receptors expressed on CT/hR-l could be tested by blockage of IL-2 binding to the endogenous murine IL-2 receptor by the anti-mouse IL-2 receptor antibody. Both murine and human IL-2 receptors are functionally active in CT/hR-l cells[39]. Hatakeyama et al.[40] have shown that human IL-2 receptors expressed by cDNA transfection of a murine T cell line, EL-4, that grows without IL-2 and does not express the murine IL-2 receptor, contain the high-affinity species. Robb[41] found that murine low-affinity receptors expressed on a non-lymphoid cell were converted to high affinity receptors by fusion of the cell membranes with human T cells. These results taken together make it reasonable to assume that high- and low-affinity IL-2 receptors arise from different states of a single receptor protein, and that a second lymphocyte-specific molecule is required for growth signal transmission and conversion of the receptor affinity from low to high[39–41].

Further studies of CT/hR-l cells gave several observations that contradicted the assumption that the IL-2 receptor forms a stable complex with a putative effector protein termed a "converter" which is essential for signal transmission and affinity change. These studies led us to suggest that the high-affinity state of the IL-2 receptor results from the formation of a ternary complex consisting of IL-2, the IL-2 receptor, and the converter.

The Number of High-Affinity Sites Is not Fixed

Based on several reports[38-41], we assumed that a complex of the IL-2 receptor and the converter protein would constitute functionally active IL-2 receptors with high affinity. A free IL-2 receptor would be functionally inactive although it binds IL-2 with low affinity. The total number of high-affinity sites of CT/hR-1 is the same or less than that of the parental line CTLL-2[39], so the number of converter molecules is limited and controlled independently from the number of IL-2 receptors. At first we thought that CT/hR-1 had a fixed number of high-affinity sites for both human and murine receptors.

We estimated the number of high- and low-affinity species of the human IL-2 receptor in CT/hR-1 by blocking IL-2 binding to the murine receptor with the monoclonal antibody PC61 as shown in Fig. 1 A. The number of human low-affinity sites (2.8×10^5 molecules per cell) was about half of the total (human plus murine) low-affinity receptors (5.2×10^5 molecules per cell) in CT/hR-1, in agreement with earlier results[39]. However, the total number of high affinity sites (3.1×10^3 molecules per cell) was almost the same regardless of the presence or absence of the monoclonal antibody, PC61, against the murine IL-2 receptor. The results show that the number of human high-affinity receptors measured by blockage of the murine receptors with antibody is the same as that of the total of both kinds of high-affinity receptor. The

result might mean that the number of high-affinity species of the murine IL-2 receptor is negligible in CT/hR-1, but that does not agree with CT/hR-1 growing in the presence of murine or rat IL-2. Note that mouse or rat IL-2 does not interact with the human IL-2 receptor although human IL-2 interacts with both human and mouse receptors.

In contrast, the presence of a small amount of mouse recombinant IL-2 drastically reduced the number of human high-affinity sites in CT/hR-1 (Fig. 3 A). The number of low-affinity sites was not much affected because the amount of murine IL-2 was not sufficient to compete with the ^{125}I-labeled human IL-2. In a control experiment, the addition of murine IL-2 did not affect the number of high-affinity sites in ATL-2, a human T-cell line derived from T cells of a patient with adult T-cell leukemia (Fig. 3 B). The result indicates that the human high-affinity receptors in CT/hR-1 disappeared when the murine high-affinity receptors were already bound by the ligand.

Affinity Conversion Model

These results are incompatible with the suggestion that a stable complex of the IL-2 receptor and the converter acts as a high-affinity receptor, which suggestion assumes a fixed number of high-affinity sites for both human and murine receptors in CT/hR-1. The findings are easily explained by an alternative suggestion, which we call the "affinity conversion model", that the converter is unable to form a complex with the IL-2 receptor unless IL-2 is bound to the receptor. This model also presumes that the apparent affinity of the

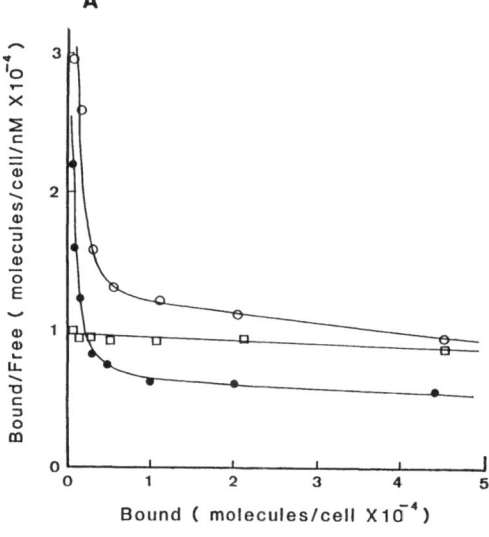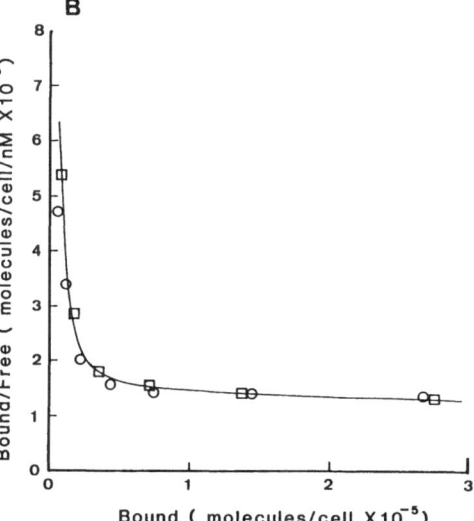

Fig. 3. Scatchard plot analysis of equilibrium binding of recombinant human ^{125}I-IL-2 to CT/hR-1 (A) and ATL-2 (B). IL-2 binding was measured in the presence of 2 mg/ml PC61 (●), 370 units/ml mouse IL-2 (□), or neither (○). With 2 mg/ml PC61, IL-2 binding to CTLL-2 was not detected (data not shown)

124

T. Kinashi *et al.*: Growth Factors and Receptors of Lymphocytes

receptor for the ligand would become high after formation of the ternary complex of the IL-2 receptor, IL-2, and the converter. As schematically shown in Fig. 4, the ligand binding to the receptor would cause a conformational change in the receptor that would increase its affinity to the converter. Ternary complex formation would, in turn, convert the apparent affinity of the receptor to the ligand from low to high. The increase in the apparent affinity to the ligand would involve an extremely low dissociation constant of the ligand from the ternary complex. In fact, the affinity of the receptor estimated by Schatchard analysis reflects the dissociation constant of the ligand from the receptor. Signal transduction may be mediated by aggregation or internalization of the ternary complex. Aggregation of the complex would be facilitated by complex formation, assuming that the converter has multiple binding sites for the receptor-ligand complex.

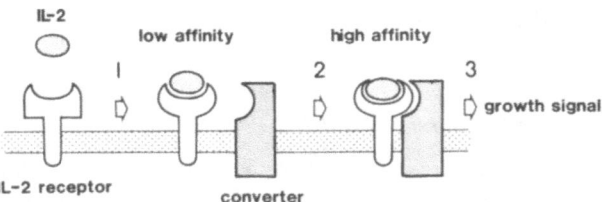

Fig. 4. Scheme of the affinity conversion model. The IL-2 receptor and converter have their top surfaces outside of the cell membrane (dotted horizontal bar). Blocking antibodies are: *1* AMT 13, PC 61, and anti-Tac; *2* HIEI; and *3* 7D4

When some of the murine receptor molecules of CT/hR-1 are occupied by murine IL-2, all of the converter molecules form ternary complexes with the murine IL-2 receptor because the number of converter molecules is less. When no converters are free, the high-affinity species of human IL-2 receptors seem to be absent. When the binding of IL-2 to murine receptors is blocked completely by antibodies, all of the converters are available for ternary complex formation with human receptors. Thus, the number of high-affinity species of the human receptor is almost equal to the number of the converter, which decides the total number of high-affinity sites.

Acknowledgements

We thank our collaborators; Drs. E. Severinson, P. Sideras, Bergstedt-Lindqvist in Stockholm, and K. Takatsu, A. Tominaga, N. Harada in Kumamoto. We thank Ms. K. Hirano for preparation of the manuscript. This investigation was supported by grants from the Ministry of Education, Science and Culture of Japan.

References

1. Honjo T (1983) Immunoglobulin genes. Ann Rev Immunol 1: 499–528
2. Honjo T, Habu S (1985) Origin of immune diversity: genetic variation and selection. Ann Rev Biochem 54: 803–830
3. Honjo T (1984) Selection theory for diversity of living organisms. Kagaku 54: 324–331, 495–502
4. Giulian D, Lachman LB (1985) Interleukin-1 stimulation of astroglial proliferation after brain injury. Science 228: 497–499
5. Benveniste EN, Merill JE (1986) Stimulation of oligodendroglial proliferation and maturation by interleukin-2. Nature 321: 610–613
6. Noma Y, Sideras P, Naito T et al (1986) Cloning of cDNA encoding the murine IgG₁ induction factor by a novel strategy using SP 6 promoter. Nature 319: 640–646
7. Kinashi T, Harada N, Severinson E et al (1987) Cloning of complementary DNA encoding T-cell replacing factor and identity with B-cell growth factor II. Nature (in press)
8. Kondo S, Shimizu A, Saito Y et al (1987) Molecular basis for two different affinity states of the interleukin 2 receptor: Affinity conversion model. Proc Natl Acad Sci USA (in press)
9. Dutton RW, Folkoff R, Hirst JA (1971) Is there evidence for a non-antigen specific diffusible chemical mediator from the thymus-derived cell in the initiation of the immune response? Prog Immunol 1: 355–368
10. Swain SL, Howard M, Kappler J et al (1983) Evidence for two distinct classes of murine B cell growth factors with activities in different functional assays. J Exp Med 158: 822–835
11. Howard M, Nakanishi K, Paul EW (1984) B cell growth and differentiation factors. Immunol Rev 78: 185–210
12. Kishimoto T (1985) Factors affecting B-cell growth and differentiation. Annu Rev Immunol 3: 133–157
13. Isakson PC, Pure E, Vitetta ES, Krammer PH (1982) T cell-derived B cell differentiation factor(s). Effect on the isotype switch of murine B cells. J Exp Med 155: 734–748
14. Bergstedt-Lindqvist S, Sideras P, MacDonald HR et al (1984) Regulation of Ig class secretion by soluble products of certain T-cell lines. Immunol Rev 78: 25–50
15. Lee F, Yokota T, Otsuka T et al (1986) Isolation and characterization of a mouse interleukin cDNA clone that expresses B-cell stimulatory factor I activities and T-cell- and mast-cell-stimulating activities. Proc Natl Acad Sci USA 83: 2061–2065
16. Schimpl A, Wecker E (1972) Replacement of T cell function by a T cell product. Nature NB 237: 15–17
17. Takatsu K, Tominaga A, Hamaoka T (1980) Antigen-induced T cell-replacing factor (TRF). I. Functional characterization of helper T lymphocytes and genetic analysis of TRF production. J Immunol 124: 2414–2422
18. Takatsu K, Harada N, Hara Y et al (1985) Purification and physico-chemical characterization of murine T cell-replacing factor (TRF). J Immunol 134: 382–389
19. Melton DA, Krieg PA, Rebagliati MR et al (1984) Efficient *in vitro* synthesis of biologically active RNA and RNA hybridization probes from plasmids containing a bacteriophage SP 6 promoter. Nucleic Acids Res 12: 7035–7056
20. Krieg PA, Melton DA (1984) Functional messenger RNAs are produced by SP 6 promoter *in vitro* transcription of cloned cDNAs. Nucleic Acids Res 12: 7057–7070
21. Okayama H, Berg P (1983) A cDNA cloning vector that permits expression of cDNA inserts in mammalian cells. Mol Cell Biol 3: 280–289

22. Yokota T, Lee F, Rennick D et al (1984) Isolation and characterization of a mouse cDNA clone that expresses mast-cell growth-factor activity in monkey cells. Proc Natl Acad Sci USA 81: 1070–1074

23. Sideras P, Bergstedt-Lindqvist S, Severinson E (1985) Secretion of IgG₁ induction factor by T cell clones and hybridomas. Eur J Immunol 15: 593–598

24. Howard M, Paul WE (1983) Regulation of B cell growth and differentiation by soluble factors. Annu Rev Immunol 1: 307–333

25. Noelle R, Krammer PH, Ohara J et al (1984) Increased expression of I a antigen on resting B cells: A new role for B cell growth factor. Proc Natl Acad Sci USA 81: 6149–6153

26. Morgan DA, Ruscetti FW, Gallo RC (1976) Selective *in vitro* growth of T lymphocytes from normal human bone marrows. Science 193: 1007–1008

27. Smith KA (1980) Interleukin 2. Immunol Rev 51: 337–353

28. Yachie A, Miyawaki T, Uwadana N et al (1983) Sequential expression of T cell activation (Tac) antigen and I a determinanta on circulating human T cells after immunization with tetanus toxoid. J Immunol 131: 731–735

29. Helmer ME, Brenner MB, McLean JM et al (1984) Antigenic stimulation regulates the level of expression of interleukin 2 receptor on human T cells. Proc Natl Acad Sci USA 81: 2171–2175

30. Robb RJ, Greene WC, Rusk CM (1984) Low and high affinity cellular receptors for interleukin 2: Implication for the level of Tac antigen. J Exp Med 160: 1126–1146

31. Robb RJ, Munck A, Smith KA (1981) T cell growth factor receptors: quantitation, specificity, and biological relevance. J Exp Med 154: 1455–1474

32. Leonard WJ, Deeper JM, Crabtree GR et al (1984) Molecular cloning and expression of cDNAs for the human interleukin-2 receptor. Nature 311: 626–631

33. Nikaido T, Shimizu A, Ishida N et al (1984) Molecular cloning of cDNA encoding human interleukin-2 receptor. Nature 311: 631–635

34. Cosman D, Cerretti PO, Larsen A et al (1984) Cloning, sequence and expression of human interleukin-2 receptor. Nature 312: 768–771

35. Shimizu A, Kondo S, Takeda S et al (1985) Nucleotide sequence of mouse IL-2 receptor cDNA and its comparison with human IL-2 receptor sequence. Nucleic Acids Res 13: 1505–1516

36. Miller J, Malek TR, Leonard WJ et al (1985) Nucleotide sequence and expression of a mouse interleukin 2 receptor cDNA. J Immunol 134: 4212–4215

37. Sabe H, Kondo S, Shimizu A et al (1984) Properties of human interleukin-2 receptors expressed on non-lymphoid cells by cDNA transfection. Mol Biol Med 2: 379–396

38. Greene WC, Robb RJ, Svetlik PB et al (1985) Stable expression of cDNA encoding the human interleukin-2 receptor in eukaryotic cells. J Exp Med 162: 363–368

39. Kondo S, Shimizu A, Maeda M et al (1986) Expression of functional human interleukin-2 receptor in mouse T cells by cDNA transfection. Nature 320: 75–77

40. Hatakeyama M, Matsumoto S, Uchiyama T et al (1985) Reconstitution of functional receptor for human interleukin-2 in mouse cells. Nature 318: 467–469

41. Robb RJ (1986) Conversion of low-affinity interleukin 2 receptors to a high-affinity state following fusion of cell membranes. Proc Natl Acad Sci USA 83: 3992–3996

Correspondence: T. Kinashi, M.D., Department of Medical Chemistry, Kyoto University Faculty of Medicine, Yoshidakonoecho, Sakyo-ku, Kyoto 606, Japan.

The book gives a survey of the medical, philosophical and religious aspects of chronic pain and suffering. Experts in the fields of neurophysiology, neuropharmacology, anaesthesiology, psychology and psychotherapy, neurology and neurosurgery as well as representatives of the main world religions and of different philosophical directions were brought together during the First Convention of the Academia Eurasiana Neurochirurgica in September 1985, and discussed the various aspects of pain and suffering, including the possibilities for treatment. The combination of religious, philosophical and medical facets of pain means a new approach to a better understanding of the problems related to pain and suffering.

Pain
A Medical and Anthropological Challenge

Proceedings of the First Convention of the Academia Eurasiana Neurochirurgica, Bonn, September 25–28, 1985

Edited by
J. Brihaye, F. Loew, H. W. Pia

Acta Neurochirurgica
Supplementum 38

1987. 111 partly coloured figures. VII, 199 pages. Cloth DM 225,–, öS 1580,– Reduced price for subscribers to "Acta Neurochirurgica": Cloth DM 202,50, öS 1422,– ISBN 3-211-81990-8

Advances in Stereotactic and Functional Neurosurgery 7

Proceedings of the 7th Meeting of the European Society for Stereotactic and Functional Neurosurgery, Birmingham 1986

Editors: Jan Gybels, Leuven, Edward R. Hitchcock, Smethwick, West Midland, Björn Meyerson, Stockholm, Christoph Ostertag, Homburg/Saar, Gian Franco Rossi, Roma

Acta Neurochirurgica
Supplementum 39

Contents: Stereotactic Imaging, Tumours and Haematomas. – Movement Disorder and Spasticity. – Pain and Miscelllaneous.

The increasing importance of stereotactic surgery in the management of common neurological conditions is illustrated by the broad applications of stereotactic techniques described in this book. International authorities present their most up-to-date experience in the fields of movement disorder, tumours, epilepsy, and pain and spasticity.

The integration of modern imaging techniques with stereotactic instrumentation is a particular feature. The book provides the most modern description of the techniques and applications of an expanding field of neurosurgery.

1987. 121 figures. IX, 199 pages. Cloth DM 225,–, öS 1580,– Reduced price for subscribers to "Acta Neurochirurgica": Cloth DM 202,50, öS 1422,– ISBN 3-211-81991-6

SPRINGER-VERLAG WIEN NEW YORK

Moelkerbastei 5, A-1010 Wien ● Heidelberger Platz 3, D-1000 Berlin 33 ● 175 Fifth Avenue, New York, NY 10010, U.S.A. ● 37-3, Hongo 3-chome, Bunkyo-ku, Tokyo 113, Japan

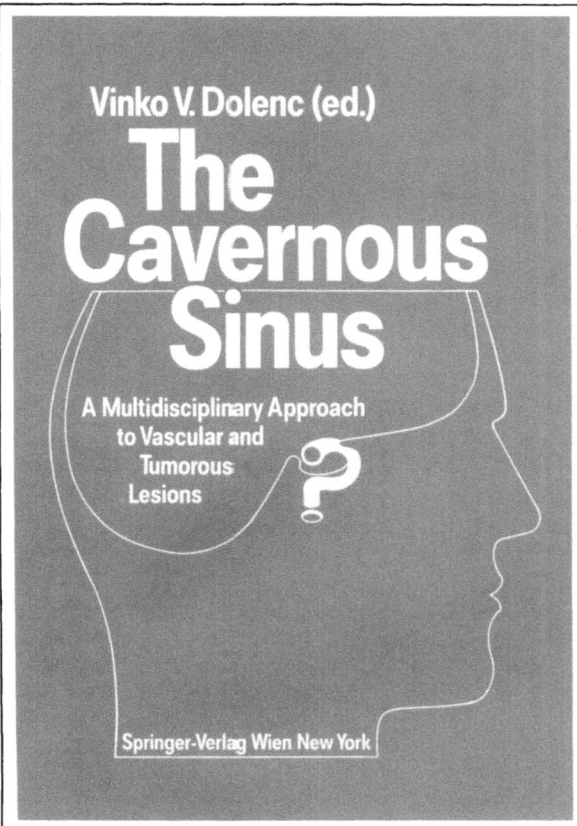